Mat
for
Soil Scientists

Join us on the web at

agriculture.delmar.cengage.com

Math
for
Soil Scientists

Mark S. Coyne
University of Kentucky

and

James A. Thompson
West Virginia University

DELMAR
CENGAGE Learning

Australia • Brazil • Japan • Korea • Mexico • Singapore • Spain • United Kingdom • United States

Math for Soil Scientists

Mark S. Coyne, James A. Thompson

Vice President, Career Education Strategic Business Unit: Dawn Gerrain

Director of Editorial: Sherry Gomoll

Acquisitions Editor: David Rosenbaum

Developmental Editor: Gerald O'Malley

Editorial Assistant: Christina Gifford

Director of Production: Wendy A. Troeger

Production Manager: J.P. Henkel

Production Editor: Kathryn B. Kucharek

Technology Project Manager: Sandy Charette

Director of Marketing: Wendy Mapstone

Marketing Specialist: Gerard McAvey

Marketing Coordinator: Erica Conley

Cover Images: USDA

Cover Design: TDB Publishing Services

For product information and technology assistance, contact us at
Cengage Learning Customer & Sales Support, 1-800-354-9706
For permission to use material from this text or product,
submit all requests online at **www.cengage.com/permissions**
Further permissions questions can be emailed to
permissionrequest@cengage.com

Library of Congress Control Number: 2004062909

ISBN-13: 978-0-7668-4268-7

ISBN-10: 0-7668-4268-1

Delmar
Executive Woods
5 Maxwell Drive
Clifton Park, NY 12065
USA

Cengage Learning is a leading provider of customized learning solutions with office locations around the globe, including Singapore, the United Kingdom, Australia, Mexico, Brazil, and Japan. Locate your local office at **www.cengage.com/global**

Cengage Learning products are represented in Canada by Nelson Education, Ltd.

To learn more about Delmar, visit **www.cengage.com/delmar**

Purchase any of our products at your local college store or at our preferred online store **www.cengagebrain.com**

Notice to the Reader

Publisher does not warrant or guarantee any of the products described herein or perform any independent analysis in connection with any of the product information contained herein. Publisher does not assume, and expressly disclaims, any obligation to obtain and include information other than that provided to it by the manufacturer. The reader is expressly warned to consider and adopt all safety precautions that might be indicated by the activities described herein and to avoid all potential hazards. By following the instructions contained herein, the reader willingly assumes all risks in connection with such instructions. The publisher makes no representations or warranties of any kind, including but not limited to, the warranties of fitness for particular purpose or merchantability, nor are any such representations implied with respect to the material set forth herein, and the publisher takes no responsibility with respect to such material. The publisher shall not be liable for any special, consequential, or exemplary damages resulting, in whole or part, from the readers' use of, or reliance upon, this material.

Printed in the United States of America
2 3 4 5 6   17 16 15 14 13

To
G. W. Thomas, R. Phillips, and V. P. Evangelou

Contents

Preface

Soil science integrates diverse fields such as chemistry, biochemistry, biology, conservation and management, fertility, morphology, and physics. They all share soil as the environment in which they occur. Science uses math, so a student or professional must master basic mathematical relationships to successfully work in soil science. The purpose of this workbook is to develop your understanding of mathematical relationships in soil science.

There are many mathematical relationships a soil scientist might use, which range from calculating the particle distribution in soil to assessing how solutes are transported through intact soil. While you can describe what is going on in soil, predicting what will happen requires a mathematical approach. The power of understanding fundamental mathematical concepts is that it lets you extrapolate to many different soil systems and to describe those systems in simple mathematical terms.

This workbook is the outgrowth of many years experience helping students investigate soil science from descriptive and mathematical perspectives. It contains problems soil science students and professional soil scientists might commonly encounter. The examples are drawn from ordinary situations that would require you to employ problem-solving skills. Each chapter begins with an overview of the math skill's importance. The use of each math skill is subsequently illustrated by an example. At the end of each chapter is a set of practice problems for which answers are provided in an accompanying teacher's guide.

We hope that by working through the examples in the workbook, you will develop confidence in your math skills, and be able to continue using a rigorous math-based approach to working in soil science and other physical sciences.

Section I

The Basics

1

Basic Math: Scientific Notation, Exponents, and Logarithms

OBJECTIVE

There are some basic math skills that are invaluable in soil science. If you are familiar with them, you can quickly move to other chapters. You may need a quick refresher, in which case these next few chapters will be good for you to review. Here are some questions you'll be able to answer by the time you finish this chapter.

- What are the benefits of using exponents and scientific notation?
- How do you convert numbers into and out of scientific notation?
- What are logarithms and how do you work with them?

Overview

Exponents and scientific notation are common in many fields of math and science, and soil science is no exception. You may wonder why we bother using scientific notation. Well, scientific notation makes it much easier to express very large or very small numbers in a more compact form. For example, it is much easier to write:

$$6.02 \times 10^{23}$$

than it is to express this number in standard format:

$$602,000,000,000,000,000,000,000$$

Not only is it quicker and easier to say, "six point zero two times ten to the twenty-third," but it is also more compact to write, and it is also less likely that you will make a mistake recording the number of zeros after the 2. In this case, there are 21 zeros. However, if you wrote 20 or 22, you would end up with a significant error at the end of your calculations.

In scientific notation, very large or very small numbers are expressed as some coefficient (an integer) times 10 to the power of some exponent.

$$\text{coefficient} \times 10^{\text{exponent}}$$

So, in scientific notation, 1,500,000 kg becomes 1.5×10^6 kg (1.5 times 10 to the power of 6). Of course, calculators (and the abacus and slide rule before that—ancient technology for some of you) make working with large numbers much easier. Imagine calculating $4382.35986 \times 309,873,558,620$ without using a calculator. You could do it, but it wouldn't be much fun. Multiplying 43.82×309.8 is easier, but might still take you some time.

Before these time saving devices, scientists resorted to other methods to make handling large or awkward numbers practical. Logarithms (logs) are basically the inverse of exponents. They are the power (exponent) to which 10 must be raised to give a particular number. The logarithm (log) of 100 = 2 because:

$$100 = 10 \times 10$$

$$100 = 10^2$$

By adding and subtracting these exponents, instead of the actual numbers, calculations could be made more quickly. The logarithms of various numbers were once printed in huge tables that scientists could consult for their calculations.

Calculations with logarithms aren't as important today as they used to be. But logarithms are still important as a means of expressing some numbers. For example, pH is a measure of the acidity in an environment. A pH of 7 is neutral. By definition, pH really means the negative logarithm of the hydrogen-ion concentration in an environment (the hydronium ion, H_3O^+, to be more accurate). This means that when pH is 7, the hydrogen-ion concentration is 10^{-7} Molar (0.0000007 M) because the log of 0.0000007 = −7. We'll talk more about concentration units later.

The logarithm isn't always an exponent of 10. The natural logarithm is used in many biological applications. The natural logarithm is the exponent of an irrational number (2.71828 . . .), the so-called natural base (base e as opposed to base 10). The principles are the same, but the notation is a little different. Natural logs are written as ln y, where y represents an integer. So, the natural logarithm (ln) of 100 = 4.605 because:

$$e^{4.605} = 100$$

$$(2.71828\ldots^{4.605} = 100)$$

We'll talk about these concepts in more detail in the next sections.

Scientific Notation

Let's review. In scientific notation, very large or very small numbers are expressed as some coefficient (usually a number between 1 and 10) times 10 to the power of some exponent. The power of an exponent is the number of times you have to multiply a number by itself. So, $100 = 10^2$ because $10 \times 10 = 100$; $1000 = 10^3$ because $10 \times 10 \times 10 = 1000$; $10,000 = 10^4$ because $10 \times 10 \times 10 \times 10 = 10,000$ (Table 1–1).

Because very big or very small numbers can have many zeros (either before or after the decimal point) scientific notation is a convenient way to write these numbers, this is done by using the powers of 10. For a big number, such as 2000, the important digit is the first one, 2. The three zeros are merely placeholders. Another way to express the number 2000 (two thousand) could be 2×1000 (two times one thousand). We also know that 1000 is $10 \times 10 \times 10$, or 10^3. Therefore, we can take 2000 and shorten it by getting rid of the zeros, but making note of how many zeros were there. The 3 then becomes the exponent in the scientific notation:

$$2000 = 2 \times 1000 = 2 \times 10^3$$

Remember, positive exponents in scientific notation represent big numbers.

Small numbers, such as 0.006, can also be represented in scientific notation using negative exponents. Recall that the negative exponent indicates division (or the denominator in a fraction). Returning then to the powers of 10, 0.006 (six thousandths) can also be expressed as $6 \div 1000$ (six divided by one thousand). As before, we know that 1000 is $10 \times 10 \times 10$, or 10^3. But because we're dividing by 1000 (not multiplying), the exponent is negative. Therefore, we can take 0.006 and shorten it by getting rid of the zeros. The -3 then becomes the exponent in the scientific notation:

$$0.006 = 6 \div 1000 = 6 \times 10^{-3}$$

TABLE 1–1 Scientific Notation

$10^5 = 10 \times 10 \times 10 \times 10 \times 10$	$= 100,000$
$10^4 = 10 \times 10 \times 10 \times 10$	$= 10,000$
$10^3 = 10 \times 10 \times 10$	$= 1000$
$10^2 = 10 \times 10$	$= 100$
$10^1 = 10$	$= 10$
10^0	$= 1$
$10^{-1} = 0.1$	$= 0.1$
$10^{-2} = 0.1 \times 0.1$	$= 0.01$
$10^{-3} = 0.1 \times 0.1 \times 0.1$	$= 0.001$
$10^{-4} = 0.1 \times 0.1 \times 0.1 \times 0.1$	$= 0.0001$
$10^{-5} = 0.1 \times 0.1 \times 0.1 \times 0.1 \times 0.1$	$= 0.00001$

To express any number in scientific notation, write it as:

$$y = b \times 10^n$$

where y is the number you're trying to turn into scientific notation, b is a number (coefficient) between 1 and 10, and n is either a positive or negative integer (e.g., $1, -1, 2, -2$). To find n, count the number of places that the decimal point must be moved to give the coefficient b. If the decimal point must be moved to the left, n = a positive integer. If the decimal point moves to the right, n = a negative integer. If no decimal point is given, you can assume it lies at the end of the right-most digit. (WARNING: *This formula only works if you are using base 10, the decimal system, but you'll be using that most of the time anyway.*)

Example 1–1

Convert 1,500,000 into scientific notation.

Solution

$$1{,}500{,}000 = b \times 10^n$$

$$1{,}500{,}000 = 1.5 \times 10^6$$

Decimal moved to the left six places.

Example 1–2

Convert 0.000000000005 g into scientific notation.

Solution

$$0.000000000005 \text{ g} = b \times 10^n \text{ g}$$

$$0.000000000005 \text{ g} = 5.0 \times 10^{-12} \text{ g}$$

Decimal moved to the right 12 places.

Example 1–3

Is 2,000,000 kg $= 200 \times 10^4$ kg?

Solution

Yes, because 2,000,000 $= b \times 10^n$

$$2{,}000{,}000 = 200 \times 10^4$$

Decimal moved to the left four places.

Exponential Notation

Exponential notation is a more general form of what we've been calling scientific notation. Exponents are also commonly used with units of measure. Common units of area include:

$$in^2 \text{ (square inches)}$$

$$m^2 \text{ (square meters)}$$

$$mi^2 \text{ (square miles)}$$

Similarly, units of volume include:

$$in^3 \text{ (cubic inches)}$$

$$cm^3 \text{ (cubic centimeters)}$$

$$m^3 \text{ (cubic meters)}$$

Exponents are also used when the unit is in the denominator of a fraction. For example, the common unit of velocity, miles per hour, could also be written as "mi h^{-1}." The negative exponent indicates that miles are divided by hours (miles per hour). Other common measures with negative exponents in the units are density, with units such as:

$$g\ cm^{-3} \text{ (grams per cubic centimeter)}$$

and application rates, such as:

$$lb\ ac^{-1} \text{ (pounds per acre)}$$

As we saw before, an exponent is really just a shorthand way to indicate repeated multiplication. For example, if you wanted to multiply the number 2 by itself five times, you could write:

$$2 \times 2 \times 2 \times 2 \times 2$$

or you could use exponents to write it as:

$$2^5 \text{ (two raised to the fifth power)}$$

This is a much more compact way of writing this operation. Either way, the result is 32.

We don't have to use a base of 10 for calculations, we might just as easily use a base of 2, 8, or 16 (these bases are particularly important in computer science). To make our formula for scientific notation more general, we can write it like this:

$$y = b \times c^n$$

where y is the number you're trying to calculate, b is a number (coefficient) usually between 1 and 10, n is the exponent, either a positive or negative integer (e.g., 1, −1, 2, −2), and c is the base you're using to express your calculations. Some of the most important calculations in soil microbiology use base 2. For example, to calculate the ideal growth of a bacterium in culture we calculate the population by using the equation for binary fission:

$$y = b \times 2^n$$

where y is the final bacterial population, b is the starting bacterial population, and n is the number of generations a bacterium grows. Notice the similarity in the formulas. For now, just recognize that in exponential notation, the exponent is the number of times you have to multiply a base by itself, as Table 1–2 illustrates.

Positive exponents are fairly straightforward in their application. However, using negative exponents is not initially as clear. A negative exponent, for example, 3^{-1}, just indicates that the number (coefficient), in this case 3,

TABLE 1–2 Exponential Notation

$2^3 = 2 \times 2 \times 2$	$= 8$	
$2^2 = 2 \times 2$	$= 4$	
$2^1 = 2$	$= 2$	
2^0	$= 1$	
$2^{-1} = 1/2$	$= 1/2$	$= 0.5$
$2^{-2} = 1/2 \times 1/2$	$= 1/4$	$= 0.25$
$2^{-3} = 1/2 \times 1/2 \times 1/2$	$= 1/8$	$= 0.125$
$4^3 = 4 \times 4 \times 4$	$= 64$	
$4^2 = 4 \times 4$	$= 16$	
$4^1 = 4$	$= 4$	
4^0	$= 1$	
$4^{-1} = 1/4$	$= 1/4$	$= 0.25$
$4^{-2} = 1/4 \times 1/4$	$= 1/16$	$= 0.625$
$4^{-3} = 1/4 \times 1/4 \times 1/4$	$= 1/64$	$= 0.015625$
$5^3 = 5 \times 5 \times 5$	$= 125$	
$5^2 = 5 \times 5$	$= 25$	
$5^1 = 5$	$= 5$	
5^0	$= 1$	
$5^{-1} = 1/5$	$= 1/5$	$= 0.2$
$5^{-2} = 1/5 \times 1/5$	$= 1/25$	$= 0.04$
$5^{-3} = 1/5 \times 1/5 \times 1/5$	$= 1/125$	$= 0.008$

is the denominator (the bottom) of a fraction (with 1 on the top):

$$3^{-1} = 1/3$$

Negative exponents are really just a way to indicate repeated divisions. More commonly, though, they are thought of as repeated multiplications of fractions. So, 4^{-3} could be expressed as:

$$1 \div 4 \div 4 \div 4 = 1 \div 64 = 0.015625$$

or

$$1/4 \times 1/4 \times 1/4 = 1/64 = 0.015625$$

Example 1–4

What is the numerical value of 2^7?

Solution

$$2^7 = 2 \times 2 \times 2 \times 2 \times 2 \times 2 \times 2$$
$$2^7 = 128$$

Example 1–5

What is the numerical value of 6^4?

Solution

$$6^4 = 6 \times 6 \times 6 \times 6$$
$$6^4 = 1296$$

Example 1–6

Which is bigger, 2^{-4} or 4^{-3}?

Solution

$$2^{-4} = 1/2 \times 1/2 \times 1/2 \times 1/2 = 1/16 = 0.0625$$
$$4^{-3} = 1/4 \times 1/4 \times 1/4 = 1/64 = 0.015625$$
$$2^{-4} > 4^{-3}$$

Exponential Numbers: Addition, Subtraction, Multiplication, and Division

There are some simple rules to follow when adding, subtracting, multiplying, and dividing numbers in scientific or exponential notation.

Addition

If both exponents are the same, simply add the coefficients while keeping the exponent term in your final answer. If exponents differ, first convert one number so that both exponents are the same and then add the coefficients (again, keeping the exponents in the final answer), or find the numerical value of each exponential number separately and add the results.

$$(b \times c^n) + (d \times c^n) = (b + d) \times c^n$$

Example 1–7

What is the sum of $(2 \times 10^2) + (4 \times 10^2)$?

Solution

$$2 \times 10^2 = 200$$

$$4 \times 10^2 = 400$$

$$200 + 400 = 600$$

or

$$(2 + 4) \times 10^2 = 6 \times 10^2 = 600$$

Example 1–8

What is the sum of $(3 \times 10^3) + (5 \times 10^2)$?

Solution

$$3 \times 10^3 = 3000$$

$$5 \times 10^2 = 500$$

$$3000 + 500 = 3500$$

or

$$3 \times 10^3 = 30 \times 10^2$$

$$(30 + 5) \times 10^2 = 35 \times 10^2 = 3500$$

Subtraction

If both exponents are the same, simply subtract the coefficients. If the exponents differ, first convert one number so that both exponents are the same and then subtract the coefficients, or find the numerical value of each exponential number separately and add the results.

$$(b \times c^n) - (d \times c^n) = (b - d) \times c^n$$

Example 1–9

Subtract 1×10^4 from 5×10^4.

Solution

$$5 \times 10^4 = 50,000$$

$$1 \times 10^4 = 10,000$$

$$50,000 - 10,000 = 40,000$$

or

$$(5 - 1) \times 10^4 = 4 \times 10^4 = 40,000$$

Example 1–10

Subtract 8×10^3 from 2×10^4.

Solution

$$8 \times 10^3 = 8000$$

$$2 \times 10^4 = 20,000$$

$$20,000 - 8000 = 12,000$$

or

$$8 \times 10^3 = 0.8 \times 10^4$$

$$(2 - 0.8) \times 10^4 = 1.2 \times 10^4 = 12,000$$

Multiplication

To multiply one exponential number by another, first multiply the coefficients together then add the exponents.

$$(b \times c^n) \times (d \times c^m) = (b \times d) \times c^{n+m}$$

Example 1–11

What is $(2 \times 10^4) \times (3 \times 10^2)$?

Solution

$$(2 \times 10^4) = 20,000$$
$$(3 \times 10^2) = 300$$
$$20,000 \times 300 = 6,000,000$$

or

$$(2 \times 3) \times 10^{4+2} = 6 \times 10^6 = 6,000,000$$

Example 1–12

What is $(5 \times 2^3) \times (6 \times 2^5)$?

Solution

$$(5 \times 2^3) = 40$$
$$(6 \times 2^5) = 192$$
$$40 \times 192 = 7680$$

or

$$(5 \times 6) \times 2^{3+5} = (30) \times 2^8 = 7680$$

What you can't do is multiply two exponential numbers together when they have different bases. You either have to convert each exponential number into its equivalent numerical value and then multiply the results, or you have to convert one of the exponential numbers so that it has the same base as the other exponential number.

Division

To divide one exponential number by another, divide one coefficient by the other and subtract the exponents. Again, the bases have to be the same to do this.

$$(b \times c^n) \div (d \times c^m) = (b \div d) \times c^{n-m}$$

Example 1–13

What is $(8 \times 10^4) \div (10 \times 10^2)$?

Solution

$$8 \times 10^4 = 80{,}000$$

$$10 \times 10^2 = 1000$$

$$80{,}000 \div 1000 = 80$$

or

$$(8 \div 10) \times 10^{4-2} = 0.8 \times 10^2 = 8 \times 10^1 = 80$$

Example 1–14

What is $(5 \times 10^4) \div (2 \times 10^6)$?

Solution

$$5 \times 10^4 = 50{,}000$$

$$2 \times 10^6 = 2{,}000{,}000$$

$$50{,}000 \div 2{,}000{,}000 = 0.025$$

or

$$(5 \div 2) \times 10^{4-6} = 2.5 \times 10^{-2} = 0.025$$

Example 1–15

What is $(4 \times 4^4) \div (1 \times 4^3)$?

Solution

$$4 \times 4^4 = 1024$$

$$1 \times 4^3 = 64$$

$$1024 \div 64 = 16$$

or

$$(4 \div 1) \times 4^{4-3} = 4 \times 4^1 = 16$$

Raising to a Power

Sometimes we have to raise an exponential number to some power. However, this is just an extension of what we have already learned about multiplying exponential numbers together. To raise an exponential number to any power, simply multiply the exponents together:

$$(c^n)^m = c^{nm}$$

Example 1–16

What is $(10^2)^4$?

Solution

$$(10^2)^4 = 10^2 \times 10^2 \times 10^2 \times 10^2 = 10^8$$

Example 1–17

What is $(5^2)^3$?

Solution

$$(5^2)^3 = 5^2 \times 5^2 \times 5^2 = 5^6$$

Logarithms

Logarithms are the exponents to which a base must be raised to get a particular numerical value. Two bases are commonly used, base 10 and the natural logarithm, which uses base e (approximately equal to 2.718). We use log_{10} or log when we want to indicate that we are using base 10 and log_e or ln when we want to indicate that we are using natural logarithms.

Logarithms aren't used very much any more to do simple calculations. However, they play a very important role in mathematically describing biological soil processes. Many numerical values measured in soil systems are transformed into logarithms to make it easier to plot, graph, and mathematically describe them.

Base 10 Logarithms

The log of $100 = 2$ and the log of $0.1 = -1$. That is because $10^2 = 100$ and $10^{-1} = 0.1$. Table 1–3 will illustrate this relationship further.

Numbers that aren't a power of 10 need to be looked up in tables of common logarithms or calculated using a calculator. For example, log 2 = 0.3010 and log 5 = 0.6989. Numbers less than one have negative logarithms. So, log 0.03 = -1.5228 and log 0.4 = -0.3979. A number less than zero can't be

TABLE 1–3 Logarithms

Value	Exponential Notation	Logarithm
10,000	10^4	4
1000	10^3	3
100	10^2	2
10	10^1	1
1	10^0	0
0.1	10^{-1}	−1
0.01	10^{-2}	−2
0.001	10^{-3}	−3
0.0001	10^{-4}	−4

TABLE 1–4 Mathematical Operations Involving Logarithms

Multiplication	$\log (nm) = \log n + \log m$
Division	$\log n/m = \log n - \log m$
Raising to a power	$\log c^n = n \log c$

assigned a logarithm. Mathematical operations involving logarithms are shown in Table 1–4.

Example 1–18

What is $\log (6 \times 2)$?

Solution

$$\log (6 \times 2) = \log 6 + \log 2$$

$$\log 6 = 0.7781$$

$$\log 2 = 0.3010$$

$$0.7781 + 0.3010 = 1.0791$$

or

$$\log (6 \times 2) = \log 12 = 1.0791$$

Example 1–19

What is $\log (1000/10)$?

Solution

$$\log (1000/10) = \log 1000 - \log 10$$
$$\log 1000 = 3$$
$$\log 10 = 1$$
$$3 - 1 = 2$$

or

$$\log 100 = 2$$

Example 1–20

What is $\log 2^8$?

Solution

$$\log 2^8 = 8 \times \log 2$$
$$\log 2 = 0.3010$$
$$8 \times 0.3010 = 2.4082$$

or

$$2^8 = 256$$
$$\log 256 = 2.4082$$

Natural Logarithms

Mathematical descriptions of certain biological transformations in soil are best described by using natural logarithms. Most calculators have the capacity to compute natural logarithms directly. There are also tables of natural log values, but these are seldom used. Remember that the base used in natural logarithms is the irrational number e (2.718 . . .). If this is the case, then $e^2 = (2.718)^2 = 7.3875$ and $\ln 7.3875 = 2$.

In practice, there is a very simple relationship between common logs and natural logs:

$$\ln b = 2.303 \log b$$

Example 1–21

What is the log and ln of 2?

Solution

$$\log 2 = 0.3010 = 10^{0.3010}$$

$$\ln 2 = 0.693 = e^{0.693} = 2.718^{0.693}$$

and

$$0.6931 = 2.303 \times 0.301$$

Mathematical operations involving natural logarithms are the same as shown in Table 1–4.

Sample Problems
Scientific Notation

1. Convert 12,000,000 into scientific notation.
2. Convert 0.125 into scientific notation.
3. Is $1 \times 10^4 = 100,000$?

Exponential Numbers: Addition, Subtraction, Multiplication, and Division

Solve the following problems by converting each of the exponent expressions into standard notation. Note that the order of mathematical operations is that exponents are solved first, multiplication and division second, and addition and subtraction are third. The first is an example.

Example

$5^3 + 2^{-2}$

Solution

First solve the exponents:

$$5^3 = 5 \times 5 \times 5 = 125$$

$$2^{-2} = {}^1\!/_2 \times {}^1\!/_2 = {}^1\!/_4 = 0.25$$

Then add the two results:

$$5^3 + 2^{-2} = 125 + 0.25 = 125.25$$

1. $1000 - 4^3$
2. $54 \div 6^2$
3. 54×6^{-2}

4. What is the numerical value of 3^8?
5. What is the numerical value of 5^2?
6. Which is larger, 2^{-5} or 5^{-2}?
7. What is $(1 \times 10^5) + (2 \times 10^5)$?
8. What is $8^2 + 4^2$?
9. What is $(2 \times 6^3) + (4 \times 6^3)$?
10. What is $4^5 - 3^5$?
11. What is $7^9 - 7^6$?
12. What is $(5 \times 4^2) - (3 \times 4^2)$?
13. What is $10^4 \times 10^5$?
14. What is $(0.1 \times 2^5) \times (5 \times 2^5)$?
15. What is $10^2 \div 10^3$?
16. What is $(10^5)^2$?

Logarithms

1. What is log 1000?
2. What is ln 1000?
3. What is log 5 + log 6?
4. What is $\ln 2^2$?

2

Significant Figures

OBJECTIVE

After completing this chapter you should be able to

- determine the appropriate significant figures to use in reporting measurements.
- use significant figures correctly in mathematical operations.
- be able to round numbers in a consistent manner.

Overview

Soil science can be inexact. The values you determine depend on how well you perform the measurement. This is why significant figures are important. They give you, and anyone reading your work, a sense of how exact your measurements are. For example, depending on the type of instrument you used, you could weigh a soil sample to 5.55, 5.547, or 5.5468 g. Presumably, the sample weight with the greatest number of digits is the most accurate, and you could imply that accuracy by reporting its value as 5.5468 ± 0.0001 g. The implication is that the true mass of the soil sample is somewhere between 5.5469 and 5.5467 g.

In practice, it is generally understood that the last digit reported is the unit in which uncertainty lies. Significant figures are routinely misused now that calculators are commonplace, because the calculators can generate values with far more digits than the accuracy of analytical equipment justifies. The purpose of this chapter is to review significant figures as a means of indicating the accuracy of various measurements in soil science.

Counting Significant Figures

In measuring soil mass, you would say that the given mass, 5.5468 g, has five significant figures compared to a value such as 5.55 g, which has only three significant figures. Likewise, 55.468 and 55.5 g have five and three significant figures, respectively. In this case, the position of the decimal point has no effect on the number of significant figures. What about a value such as 1,000,000? Does it have one or seven significant figures? Unfortunately, without knowing how the value was measured, it's difficult to say. If the value had been reported as 1.0×10^6, you would know immediately that it has two significant figures and that the accuracy of the measurement was probably ±100,000.

Example 2–1

How many significant figures are there in a value of 910 mL?

Solution

Two or three depending on whether the measured value reflects an accuracy of ±1 mL (three significant figures) or an accuracy of ±10 mL (two significant figures), in which case, it would be better to have reported it as 9.1×10^2 mL.

What if a value is given as 0.00246? Are there three or six significant figures? In cases such as these, the zero only fixes the position of the decimal point, and so there are only three significant figures. You can convince yourself that this is reasonable by transforming the value into scientific notation, in which case, it is 2.46×10^{-3}, and there is no ambiguity. On the other hand, 1.00246 would have six significant figures because the zeros following the decimal point are not fixing their positions.

Multiplication and Division

The general rule for multiplying and dividing two inexact numbers together is to keep the significant figures of the least accurate number.

Example 2–2

What is the number of significant figures in the product of 6.28×7.3?

Solution

6.28 has three significant figures and 7.3 has two significant figures. The product of the two values is 45.844, but it should be reported as 46.

Example 2–3

What is the number of significant figures in the quotient of $3.1426 \div 2.83$?

Solution

3.1426 has five significant figures while 2.83 has just three significant figures. 3.1426 ÷ 2.83 = 1.110. To use the appropriate number of significant figures we would report this value as 1.11.

Addition and Subtraction

Determining significant figures in addition and subtraction problems uses a slightly different principle than that used for multiplication and division. In adding or subtracting numbers, the significant figures depend on the least precise component of that equation.

Example 2–4

How do you express the total mass of a mixture that contains 14 g water, 0.007 g KCl, and 1.42 g NaCl?

Solution

In this case, the total mass of the mixture is 15 g because the measurement of water had the least accuracy. Even though the mass of KCl was reported to just one significant figure, it is clear that the instrument used was (hopefully) very precise.

Rounding Off

It is frequently necessary to round off numbers when decisions about significant figures are made. Rules for rounding off a number are as follows:

1. If the final digit dropped is less than five, do not change the preceding digit.
2. If the final digit dropped is greater than five, increase the preceding digit by one.
3. If the final digit dropped is five, increase the preceding digit by one if it is odd and leave it unchanged if it is even.

Example 2–5

Round the following numbers to two significant figures: 8.47, 8.43, 6.65, 3.75, 9.35.

Solution

$$8.47 = 8.5$$
$$8.43 = 8.4$$
$$6.65 = 6.6$$
$$3.75 = 3.8$$
$$9.35 = 9.4$$

Sample Problems

Counting Significant Figures

1. How many significant figures are there in 2.06×10^{-3} g?
2. How many significant figures are there in 0.036 mL?
3. How many significant figures are there in 0.05×10^{-2} g?
4. How many significant figures are there in 550 m?
5. How many significant figures are there in 1.0234×10^{34}?

Multiplication and Division

1. What is the value and number of significant figures in the product of $(6.10 \times 10^3) \times (2.08 \times 10^{-4})$?
2. Multiply 0.00497×211 and record your answer to the correct number of significant figures.
3. Multiply 1.0623×0.00517 and record your answer to the correct number of significant figures.
4. Carry out the following calculation and record your answer to the correct number of significant figures. Which number controls your choice of significant figures?

$$\frac{34.73 \times 0.5531 \times 0.05300}{1.2345} = ?$$

5. Carry out the following calculation and record your answer to the correct number of significant figures.

$$\frac{41.58 \times 991}{133.5 \times 0.6312} = ?$$

Addition and Subtraction

1. Add the following numbers and report the results to the correct number of significant figures.

$$5.92 + 3.1111 + 1375$$

2. Add the following numbers and report the results to the correct number of significant figures.

$$5.21 \times 10^{-3} + 1.93 + 2.20 \times 10^{-2}$$

3. Add the following numbers and report the results to the correct number of significant figures.

$$0.100 + 1.000 + 1.4$$

4. Subtract the following numbers and report the results to the correct number of significant figures.

$$215 - 0.05$$

Rounding Off

1. Round 4.3958 to two significant figures.
2. Round 98.4398 to three significant figures.
3. Round 34,987.5940 to five significant figures.
4. Round 1.5 to one significant figure.
5. Round 12.65 to four significant figures.

3

Metric and International Scientific (SI) Units

OBJECTIVE

After completing this chapter you should be able to

- move easily among metric unit scales.
- convert between English and scientific units.

Overview

Scientists use the metric system to express their results and soil scientists should be no different, although many conventions for reporting measurements still use English units. **Metric units** are ideal to work with because they are based on a decimal system (as is the U.S. currency) in which different scales of measure always differ by factors of 10. Each scale has a distinct prefix that identifies it. The names of the most common prefixes used in soil science are given in Table 3–1.

For example, 1 m (meter) (approximately 1.09 yd(yards)) might be used to express length. One meter is composed of 100 cm (centimeters) and 1 cm is composed of 10 mm (millimeters). So, 1000 mm is the same length as 1 m. You can convert between different ways of expressing the same measurement in the metric system simply by moving the decimal point.

Example 3–1

Express 92,067 μm (micrometers) in centimeters, decimeters (dm), and meters.

TABLE 3–1   Prefixes for Metric Scales and SI Units

Prefix	Relative Scale	Abbreviation	Example
tera	10^{12}	T	teragram (Tg)
giga	10^{9}	G	gigagram (Gg)
mega	10^{6}	M	megagram (Mg)
kilo	10^{3}	k	kilogram (kg)
			kilometer (km)
	10^{0}		gram (g)
			liter (L)
			meter (m)
deci	10^{-1}	d	decimeter (dm)
centi	10^{-2}	c	centimeter (cm)
milli	10^{-3}	m	milligram (mg)
			milliliter (mL)
			millimeter (mm)
micro	10^{-6}	μ	microgram (μg)
			microliter (μL)
			micrometer (μm)
nano	10^{-9}	n	nanogram (ng)
			nanoliter (nL)
			nanometer (nm)
pico	10^{-12}	p	picogram (pg)
femto	10^{-15}	f	femtogram (fg)

Solution

$$92{,}067 \ \mu\text{m}$$
$$9.2067 \ \text{cm}$$
$$0.92067 \ \text{dm}$$
$$0.090267 \ \text{m}$$

Note that each expression has five significant figures.

Systéme Internationale (SI) units represent the preferred system for expressing everything from length to yield in science. The SI units are based on metric measurements. For example, the preferred units for expressing mass are gram (g), kilogram (kg), or megagram (Mg). The preferred units for expressing length are meter and kilometer (km). The preferred units for expressing volume are liter (L) or cubic meter (m^3). Other SI units for expressing measurements in soil science will be discussed later in this chapter.

Measures of Length

The basic unit for measuring length in the metric system is the **meter,** which is slightly longer than 1 yd. Lengths much greater than 1 m are usually expressed as kilometers, and lengths much smaller than 1 m are usually expressed in centimeters, millimeters, or micrometers, depending on the appropriate scale. For example, the size of clay and other soil colloids is best described by using micrometers or nanometers (nm). Likewise, the size of most protozoa, fungi, and bacteria in soil is best described by micrometers, while viruses are in the nanometer range. The relationship between the various measures of length in relation to 1 m is illustrated below:

$$1\text{ m} = 0.0010\text{ km} \qquad (10^{-3}\text{ km m}^{-1})$$

$$1\text{ m} = 1.0\text{ m} \qquad (10^{0}\text{ m m}^{-1})$$

$$1\text{ m} = 10\text{ dm} \qquad (10^{1}\text{ dm m}^{-1})$$

$$1\text{ m} = 100\text{ cm} \qquad (10^{2}\text{ cm m}^{-1})$$

$$1\text{ m} = 1000\text{ mm} \qquad (10^{3}\text{ mm m}^{-1})$$

$$1\text{ m} = 1{,}000{,}000\ \mu\text{m} \qquad (10^{6}\ \mu\text{m m}^{-1})$$

$$1\text{ m} = 1{,}000{,}000{,}000\text{ nm} \qquad (10^{9}\text{ nm m}^{-1})$$

There are 1000 μm in 1 mm, 1000 mm in 1 m, and 1000 m in 1 km. As you have seen before, however, it is relatively easy to shift between scales simply by moving the decimal point.

Example 3–2

How many millimeters are in 1 km?

Solution

$$1\text{ km}$$

$$1000\text{ m}$$

$$1{,}000{,}000\text{ mm}$$

or

$$1\text{ km} \times (10^{3}\text{ m km}^{-1}) \times (10^{3}\text{ mm m}^{-1}) = 10^{6}\text{ mm} = 1{,}000{,}000\text{ mm}$$

Measures of Area

The typical units of area you will encounter in soil science are the square centimeter (cm^2), the square meter (10^4 cm^2), the hectare (10^4 m^2), and the square kilometer (100 ha or 10^6 m^2).

Example 3–3

What is the area of a square 1.5 cm on each side?

Solution

$$1.5 \text{ cm} \times 1.5 \text{ cm} = 2.25 \text{ cm}^2$$

Example 3–4

How many hectares is a plot measuring 3 m by 15 m?

Solution

$$3 \text{ m} \times 15 \text{ m} = 45 \text{ m}^2$$

$$45 \text{ m}^2 \times 1 \text{ ha}/10{,}000 \text{ m}^2 = 0.005 \text{ ha}$$

Measures of Volume

The basic unit for measuring volume in the metric system is the **liter**. In SI units, the liter is defined as 10^{-3} m^3. This is the equivalent volume enclosed by a cube with dimensions of 0.1 m per side (or a cube 10.0 cm on each side). A **milliliter** (mL) is equivalent to 1 cm^3. So, in a cube with 10 cm per side, the total volume would be 1000 cm^3 or 1000 mL. Other frequently used measures of volume are the **microliter** (10^{-6} L) and the **nanoliter** (10^{-9} L). In the metric system, the means of converting among scales of volume measurement are exactly as we saw with measures of length.

Example 3–5

Convert 15 mL into L and μL.

Solution

$$15 \text{ mL}$$

$$0.015 \text{ L}$$

$$15{,}000 \ \mu\text{L}$$

Example 3–6

How many mL will it take to fill a 5.5 L flask?

Solution

$$5.5 \text{ L} \times 1000 \text{ mL L}^{-1} = 5500 \text{ mL}$$

Measures of Mass

The basic unit for measuring mass in the metric system is the **gram.** There are 1000 g in 1 kg and 1000 kg in 1 Mg (also called the metric ton). Likewise, there are 1000 mg in 1 g and 1000 μg in 1 mg. As you saw with length, it's easy to convert between scales simply by moving the decimal point.

Example 3–7

Convert 13 g into kg and μg.

Solution

$$13 \text{ g}$$
$$0.013 \text{ kg}$$
$$13,000,000.0 \ \mu\text{g}$$

Measures of Concentration

Measures of solution concentration in the metric system are invariably moles per liter or equivalents per liter (**molarity** and **normality,** respectively). We'll talk about molarity and normality calculations later when we discuss soil chemistry, which is when these concepts become very important. When dealing with solids, however, several other measures of concentration are used. The most important are **parts per million** (ppm) and **parts per billion** (ppb). These are outlined in Table 3–2. Note that ppm and ppb are not SI units. We also use ppm and ppb when we use solution concentrations in soil science.

Example 3–8

If the concentration of atrazine in water is 0.1 gL^{-1}, how many ppm is that?

Solution

Parts per million is milligrams per liter

$$0.1 \text{ g} = 100 \text{ mg}$$

The atrazine concentration is 100 ppm.

TABLE 3–2 Common Units for Expressing Concentration

Unit	Wt/Wt Basis	Wt/Vol Basis	Vol/Vol Basis
Parts per million (ppm)	$mg\ kg^{-1}$	$mg\ L^{-1}$	$\mu L\ L^{-1}$
	$\mu g\ g$	$\mu g\ mL^{-1}$	$nL\ mL^{-1}$
Parts per billion (ppb)	$\mu g\ kg^{-1}$	$\mu g\ L^{-1}$	$nL\ L^{-1}$
	$ng\ g^{-1}$	$ng\ mL^{-1}$	$pL\ mL^{-1}$
Percent (%)	g/100g	g/100 mL	mL/100 mL

Example 3–9

If the maximum lead (Pb) concentration in a soil can be 1000 ppm, are you over this limit if you measure a lead concentration of 1200 $\mu g\ g^{-1}$?

Solution

$$1200\ \mu g = 1.2\ mg$$

$$1\ g = 0.001\ kg$$

1.2 mg ÷ 0.001 kg = 1200 mg kg^{-1}, which is 1200 ppm—over the limit.

Other Important SI Units of Measure

There are several other important SI units to become familiar with because they form part of the language soil scientists use in problem solving. Yields are typically reported as kg ha^{-1} or Mg ha^{-1}. Pressure is reported in terms of **megapascals** (MPa) or **pascals** (Pa). 1 MPa = 10^6 Pa.

Temperature is usually reported in terms of degrees Celsius (°C) although Kelvin (K) is used in some applications. 0 K = −273°C. The SI unit for energy is the joule (J). The unit to report radioactivity is the **becquerel** (Bq).

Sample Problems
Measures of Length

1. How many nanometers are present in 10 dm?
2. Convert 12,000 cm to kilometers and micrometers.
3. In a 1500-m race, how many micrometers are the contestants running?
4. Convert 60 km into the appropriate number of centimeters.
5. Use scientific notation to express the number of nanometers in 100 cm.

Measures of Area

1. How many square centimeters are in 1 ha?
2. How many square meters are in 0.1 ha?
3. What is the area of a plot 30.5 m wide by 35 m long? Use the appropriate number of significant figures to report your answer.
4. How many hectares are contained in a plot 100 m by 50 m?
5. How many square micrometers are in a sugar cube with dimensions of 1 cm per side?

Measures of Volume

1. How many liters are present in 1 m^3?
2. If a gas tank has a 40 L capacity, how many cm^3 does it hold?
3. How many microliters are present in 500 mL?
4. If 1.0 mL has a mass of approximately 1.0 g, what is the mass of 0.075 mL?
5. How many milliliters would it take to fill a tank with a volume of 0.01 m^3?

Measures of Mass

1. How many milligrams are present in 1 Mg?
2. How many grams are present in 10^6 ng?
3. If the surface 15 cm of soil in 1 ha typically has a mass of 2×10^6 kg, how many grams does it contain?
4. Which has the greater mass 10^{10} ng or 10^3 mg?
5. If the combined mass of three individuals is 150 kg, what is their average weight in terms of megagrams?

Measures of Concentration

1. What is the concentration of a salt solution that contains 8 g NaCl by weight if the final volume is 0.9 L?
2. What is the concentration of the pesticide metolachlor if a 500 mL solution contains 0.3 μg of metolachlor?
3. If a 5.0-g plant sample contains 3.5 μg of zinc, what is the zinc concentration in the plant tissue in ppm?
4. A soil sample is contaminated with 100 ppm toluene. What is the actual amount of toluene in a 10-g sample?
5. If there is 1 ppb arsenic in the upper 15 cm of 1 ha of soil, how much total arsenic is present in this environment?

4

Unit Conversions

OBJECTIVE

The purpose of this chapter will be to review the usefulness of the conversion factor method and provide you with a table of the most common conversions a soil scientist might use. After completing this chapter, you should be able to

- shift between English and SI units with ease.
- convert unsuitable unit expressions to values in common use.

Overview

You may have noticed that some of the problems in Chapter 3 were solved by multiplying strings of conversion factors together. This is a pedestrian but highly effective technique for making conversions between units. By multiplying strings of conversion factors together, and canceling out similar terms in the numerator and denominator, you can end up with a correct numerical answer that is also expressed in the correct units.

Unit conversions are particularly important for soil scientists because the units in which relationships are expressed are critical. There is a world of difference between soil processes that produce 10 kg of nitrogen (N) compared to 10 mg of N. Unit conversions are most critical when soil scientists have to convert between SI and non-SI units. Although soil scientists may use the metric system and SI units, they can't do so exclusively because there is a parallel system of non-SI units for mass, volume, distance, etc., that are in common use.

The Conversion Factor Method

Let's say you wanted to convert 12 in to units of feet. How would you do it? If you already know that 12 in = 1 ft, it's easy. But what if you only have 7 in? It becomes more difficult. The conversion factor method is a straightforward way to solve problems like this.

Example 4–1

Convert 7 in to feet.

Solution

$$7 \text{ in} \times \frac{1 \text{ ft}}{12 \text{ in}} = ? \text{ ft}$$

$$7 \text{ in} \times \frac{1 \text{ ft}}{12 \text{ in}} = 0.6 \text{ ft}$$

Example 4–2

How many moles of glucose are present in 17 g of glucose if the formula weight of glucose is 174 g/mol?

Solution

$$17 \text{ g glucose} \times \frac{1 \text{ mol glucose}}{174 \text{ g glucose}} = ? \text{ mol glucose}$$

$$17 \text{ g glucose} \times \frac{1 \text{ mol glucose}}{174 \text{ g glucose}} = 0.01 \text{ mol glucose}$$

Similar terms in the numerator and denominator cancel out and all you have to do is carry out a simple division.

In the previous examples, the most important thing to remember about unit conversion was that the unit you wanted to end up with needed to be in the numerator. This isn't always the case. Sometimes you will be working with unit conversions in which you have to arrive at the proper unit in both the numerator and denominator such as converting from pounds per acre (lb ac^{-1}) to kilograms per hectare (kg ha^{-1}).

Example 4–3

Convert 2 million pounds per acre to kilograms per hectare.

Solution

$$\frac{2 \times 10^6 \text{ lb}}{\text{acre}} \times \frac{0.454 \text{ kg}}{\text{lb}} \times \frac{2.47 \text{ acre}}{\text{ha}} = \frac{? \text{ kg}}{\text{ha}}$$

$$\frac{2 \times 10^6 \,\cancel{\text{lb}}}{\cancel{\text{acre}}} \times \frac{0.454 \text{ kg}}{\cancel{\text{lb}}} \times \frac{2.47 \,\cancel{\text{acre}}}{\text{ha}} = \frac{2,242,760 \text{ kg}}{\text{ha}} = \frac{2 \times 10^6 \text{ kg}}{\text{ha}}$$

Remember those pesky significant figures? Because 2 million lb acre^{-1} is the term with the fewest significant figures (one), even though our calculator can give us seven significant figures (2,242,760) we should report the result as 2×10^6 kg ha^{-1} (only one significant figure).

The string of conversion factors can be quite long, particularly if you're trying to convert between very different units.

Example 4–4

How many micrograms of N are present in 650 mL of gas that has a concentration of 5 ppm N_2O?

Solution

Remember that parts per million (ppm) = microliters per liter (μL L^{-1}). It is also useful to know (critical to know) that the universal gas conversion factor is 1 mol per 22.4 L or 1 μmol per 22.4 μL. Given that

$$5 \text{ ppm} = \frac{5 \,\cancel{\mu\text{L}}}{\cancel{\text{L}}} \times \frac{1 \,\cancel{\mu\text{mole}}}{22.4 \,\cancel{\mu\text{L}}} \times 0.65 \,\cancel{\text{L}} \times \frac{28 \,\mu\text{g N}}{\cancel{\mu\text{mole N}_2\text{O}}} = 4 \,\mu\text{g N}$$

Converting between SI and Non-SI Units

Table 4–1 is an extremely useful tool that allows you to look up the most common conversion factors to change SI units to non-SI units and vice versa. It lets you skip the step of canceling out like terms in an equation. Depending on which unit you want to convert, you simply multiply it by the conversion factor to either its right or left. For example, convert 15 km to miles.

Example 4–5

Convert 15 km into its equivalent length in miles.

Solution

$$15 \text{ km} \times 0.621 = 9.32 \text{ mi}$$

TABLE 4–1 **Conversion Factors for SI and Non-SI Units**

SI Unit	To Convert to Non-SI Unit Multiply by	Non-SI Unit	To Convert to SI Unit Multiply by
Length			
kilometer	0.621	mile	1.609
meter	1.094	yard	0.914
meter	3.28	foot	0.304
millimeter	3.94×10^{-2}	inch	25.4
nanometer	10	Angstrom	0.1
Area			
hectare	2.47	acre	0.405
square km	247	acre	4.05×10^{-3}
square km	0.386	square mile	2.590
square m	2.47×10^{-4}	acre	4.05×10^{3}
square m	10.76	square foot	9.29×10^{-2}
square mm	1.55×10^{-3}	square inch	645
Volume			
cubic meter	9.73×10^{-3}	acre-inch	102.8
cubic meter	35.3	cubic foot	2.83×10^{-2}
liter	6.10×10^{4}	cubic inch	1.64×10^{-5}
liter	2.84×10^{-2}	bushel	35.24
liter	1.057	quart	0.946
liter	3.53×10^{-2}	cubic foot	28.3
liter	0.265	gallon	3.78
liter	33.78	ounce	2.96×10^{-2}
liter	2.11	pint	0.473
Mass			
gram	2.20×10^{-3}	pound	454
gram	3.552×10^{-2}	ounce	28.4
kilogram	2.205	pound	0.454
kilogram	1.10×10^{-3}	ton	907
megagram	1.102	ton	0.907

TABLE 4–1 (Continued)

SI Unit	To Convert to Non-SI Unit Multiply by	Non-SI Unit	To Convert to SI Unit Multiply by
Yield or Rate			
kg ha^{-1}	0.893	lb acre^{-1}	1.12
kg m^{-3}	7.77×10^{-2}	lb bu^{-1}	12.87
kg ha^{-1}	1.49×10^{-2}	bu acre^{-1}	67.19
kg ha^{-1}	1.59×10^{-2}	bu acre^{-1}	62.71
kg ha^{-1}	1.86×10^{-2}	bu acre^{-1}	53.75
L ha^{-1}	0.107	gal acre^{-1}	9.35
Mg ha^{-1}	893	lb acre^{-1}	1.12×10^{-3}
Mg ha^{-1}	0.446	ton acre^{-1}	2.24
m sec^{-1}	2.24	mph	0.447
Pressure			
megapascal	9.90	atmosphere	0.101
megapascal	10	bar	0.10
pascal	2.09×10^{-2}	lb ft^{-2}	47.9
pascal	1.45×10^{-4}	lb in^{-2}	6.90×10^{3}
Temperature			
Kelvin	1(K-273)	Celsius	1(°C + 273)
Celsius	(9/5 °C) + 32	Fahrenheit	5/9(°F − 32)
Energy, Work, or Heat			
joule	9.52×10^{-4}	Btu	1.05×10^{3}
joule	0.239	calorie	4.19
joule	107	erg	10^{-7}
joule	0.735	foot-pound	1.36
joule m^{-2}	2.387×10^{-5}	cal cm^{-2}	4.19×10^{4}
newton (N)	10^{5}	dyne	10^{-5}
watt (W) m^{-2}	1.43×10^{-3}	cal cm^{-2} min^{-1}	698
Water Measurements			
m^{-3}	9.73×10^{-3}	acre-inches	102.8
m^{-3} h^{-1}	9.81×10^{-3}	ft^{3} sec^{-1}	101.9

(*continued*)

TABLE 4–1 (Continued)

SI Unit	To Convert to Non-SI Unit Multiply by	Non-SI Unit	To Convert to SI Unit Multiply by
		Water Measurements	
$m^{-3} h^{-1}$	4.40	gal min^{-1}	0.227
ha-meter	8.11	acre-feet	0.123
ha-meter	97.28	acre-inch	103×10^{-2}
ha-cm	8.1×10^{-2}	acre-feet	12.33
		Concentration	
centimole per kg	1	milliequivalent per 100 g	1
g kg^{-1}	0.1	percent	10
mg kg^{-1}	1	ppm	1
		Radiation	
becquerel (Bq)	2.77×10^{-11}	curie (Ci)	3.7×10^{10}
Bq kg^{-1}	2.7×10^{-2}	picocurie per g^{-1}	37
gray (Gy)	100	rad	0.01
sievert (Sv)		rem (roentgen equiv. man)	0.01

Note. American Society of Agronomy, Crop Science Society of America, Soil Science Society of America (ASA-CSSA-SSSA). (1998). *Publications and style manual.* Madison, WI: Author.

This is the functional equivalent of having written the following equation to solve the same problem:

$$15 \text{ km} \times \frac{0.621 \text{ mi}}{\text{km}} = ? \text{ mi}$$

$$15 \ \cancel{\text{km}} \times \frac{0.621 \text{ mi}}{\cancel{\text{km}}} = 9.315 \text{ mi}$$

Remember our rule for significant figures. In multiplication, the final answer has the same number of significant figures as the term with the least number of significant figures; 15 km has two significant figures, 0.621 mi km^{-1} has three significant figures. So we should round our answer up to 9.3 mi.

Example 4–6

Convert 15Å to nanometers.

Solution

$$15 \text{ Å} \times 0.1 = 1.5 \text{ nm}$$

Sample Problems

1. How many feet is 1500 m?
2. How many square meters are in 2 acre?
3. How many gallons are in 5 L?
4. How many kilograms are in 10 ton of coal?
5. What is the yield, in bushels per acre, of a plot that yielded 1000 kg ha^{-1}?
6. How many megapascals are in 10 atm?
7. 37°C is what temperature in degrees Fahrenheit?
8. Absolute zero is 0 K. What temperature is that in degrees Celsius and degrees Fahrenheit?
9. 110 cal represent how many joules?
10. How many acre feet of water are in 2 ha cm of water?
11. Convert 6 lb into micrograms.
12. How many centimeters are in 0.75 yd?
13. Express 15 Pa as equivalent atmospheres.
14. How many square feet are in 1 ha?
15. How many fluid pints are in 1 ft^3?

Section II

Quantifying Physical and Chemical Properties of Soils

5

Soil Texture and Surface Area

OBJECTIVE

After completing this chapter you should be able to

- determine the relative proportions of sand, silt, and clay in a soil sample from standard laboratory data.
- use a textural triangle to identify the soil textural class of a soil sample after being given particle size distribution data.
- estimate the surface area of a sample based on the particle size distribution.

Overview

There is more to soil than the mere weight of its components. To best understand the nature of soil, the processes that occur in soil, and the unique role of soil in natural and managed systems, you must be able to solve problems dealing with the special properties of the mineral and organic material in soil, and how they interact with each other as well as the water and air in the soil. We start that examination in this chapter by looking at the physical properties of soil. We will examine the mineral solids in soil, and answer several key questions about them, such as what the primary particles are that, on a weight basis, dominate the composition of most soils and how to determine texture based on some basic analytical measurements.

Soil Texture—Particle Size and Surface Area

Soil is composed of different-sized mineral particles that are grouped into three classes in the USDA system of soil classification:

Sand—Particles 0.05 to 2.0 mm in diameter

Silt—Particles 0.002 to 0.05 mm in diameter

Clay—Particles <0.002 mm in diameter

The texture of a soil is determined by the distribution of these three classes of soil particles. Only particles <2.0 mm in diameter are considered for textural analysis. Texture is important because it affects the aeration, drainage, and workability of a soil and its cation exchange capacity (CEC) and cohesiveness. To determine the percent of each particle-size fraction in a soil sample for textural analysis, divide the mass of soil in each fraction by the total mass of soil less than 2.0 mm in diameter, and multiply by 100.

Example 5–1

If 1 kg of soil contains 50 g of particles <0.002 mm in diameter, 80 g of particles >0.05 but <2.0 mm in diameter, and 150 g of particles between 0.002 and 0.05 mm in diameter, what is the percent of sand, silt, and clay in this soil?

Solution

$$50 \text{ g} + 80 \text{ g} + 150 \text{ g} = 280 \text{ g}$$

$$\text{Sand sized (0.05 to 2.0 mm): } \left(\frac{80 \text{ g}}{280 \text{ g}} \right) \times 100 = 29\%$$

$$\text{Silt sized (0.002 to 0.05 mm): } \left(\frac{150 \text{ g}}{280 \text{ g}} \right) \times 100 = 54\%$$

$$\text{Clay sized (<0.002 mm): } \left(\frac{50 \text{ g}}{280 \text{ g}} \right) \times 100 = 18\%$$

The values won't add up to exactly 100% because of rounding.

Remember that only particles <2.0 mm in diameter are used for textural class determinations. The bulk of the soil in Example 5–1 exceeded this size, which indicates that it was very coarse.

The specific surface area is the amount of surface exposed by particles per unit of weight, usually expressed in terms of $cm^2 \, g^{-1}$. As the size (i.e., diameter) of a particle decreases, its specific surface area increases. Typical specific surface areas for sand, silt, and clay-sized fractions are 30, 1500 and 3,000,000 $cm^2 \, g^{-1}$, respectively (Thein & Graveel, 1997). As the surface area increases, the interaction between soil particles and their surrounding environment also increases. Consequently, fine-textured soils, which contain large surface areas, are more chemically reactive than coarse-textured soils. Likewise, the contribution of different fractions to a soil's chemical and physical properties is not proportional to a fraction's abundance. A few clay-sized particles can have a lot more influence than many sand-sized particles.

Example 5–2

For a 12 g soil sample that contains 60 percent sand, 30 percent silt, and 10 percent clay, calculate the specific surface area of each fraction and the percent contribution of that fraction to the total surface area of the soil sample.

Solution

Step 1. Determine the mass of each fraction.

$$12 \text{ g} \times 60\% \text{ sand} = 7.2 \text{ g sand}$$

$$12 \text{ g} \times 30\% \text{ silt} = 3.6 \text{ g silt}$$

$$12 \text{ g} \times 10\% \text{ clay} = 1.2 \text{ g clay}$$

Step 2. Determine the area contributed by each fraction.

$$7.2 \text{ g sand} \times 30 \text{ cm}^2 \text{ g}^{-1} = 216 \text{ cm}^2$$

$$3.6 \text{ g silt} \times 1500 \text{ cm}^2 \text{ g}^{-1} = 5400 \text{ cm}^2$$

$$1.2 \text{ g clay} \times 3{,}000{,}000 \text{ cm}^2 \text{ g}^{-1} = 3.6 \times 10^6 \text{ cm}^2$$

Step 3. Determine the portion of total area contributed by each fraction.

$$\text{Total specific surface area} = 3{,}605{,}616 \text{ cm}^2$$

Fraction	% Contribution to specific surface area
Sand	$(216 \text{ cm}^2/3{,}605{,}616 \text{ cm}^2) \times 100 = 0.006\%$
Silt	$(5400 \text{ cm}^2/3{,}605{,}616 \text{ cm}^2) \times 100 = 0.150\%$
Clay	$(3{,}600{,}000 \text{ cm}^2/3{,}605{,}616 \text{ cm}^2) \times 100 = 99.84\%$

You can see the significance of size and surface area. Sand, which contributes 60 percent of the mass of this sample, contributes less than 0.01 percent of the surface area, while clay, which makes up only 10 percent of the mass of the sample, contributes over 99 percent of the surface area.

Soil Texture—Using the Textural Triangle

Based on the particle size distribution, mineral soils can be placed in one of 12 textural classes by using a textural triangle (Figure 5–1). Knowing the textural class of a soil is important because soil physical properties and reactive surface area are strongly influenced by the size of its particles. As the surface area of a

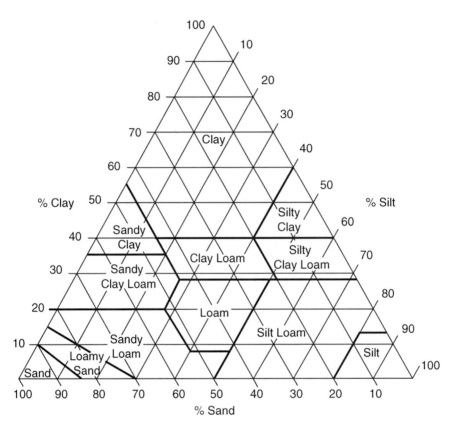

FIGURE 5–1 Soil textural triangle.

Source. Soil survey staff. (1975). Soil taxonomy. In *Agricultural handbook no. 436. soil conservation service, USDA.* Washington, DC: U.S. Govt. Printing Office.

soil or soil particle increases, its chemical and physical reactivity also increase. In addition, important features like porosity and water transport are influenced by texture. For example, the suitability of soils for the installation of on-site septic systems is partly determined by the textural class.

To use the textural triangle, first determine the percent of sand, silt, and clay. Organic soils do not fit into this classification scheme, even if you can determine their particle size distribution. The textural triangle has one side labeled for each fraction—sand, silt, or clay. For any two of the fractions, follow the line corresponding to the percent of that fraction either up, down, or across (depending on which fraction you choose). For example, to determine the clay percentages follow the lines across horizontally for clay, down and to the left for silt, and up and to the left for sand. The intersection of the two lines is in the textural class of that soil. It doesn't matter which two fractions you choose. You will arrive at the same textural class regardless of whether you use sand and silt, sand and clay, or clay and silt. As a general rule, laboratory analyses for texture typically determine the sand- and silt-sized fractions, and the clay fraction is determined by difference.

Example 5–3

For a 100 g soil sample consisting of 75 g of particles <2.0 mm in diameter, what is the mass and percent of the clay-sized particles if the sample consists of 50 percent sand and 15 percent silt?

Solution

Remember that textural analysis is only performed on fractions <2.0 mm in diameter, so we are only concerned with 75 g of soil that meet the criterion.

$$50\% \text{ sand} \times 75 \text{ g soil} = 37.5 \text{ g sand}$$

$$15\% \text{ silt} \times 75 \text{ g soil} = 11.2 \text{ g silt}$$

Percent and mass of clay is determined by difference:

$$100\% - (50\% + 15\%) = 35\%$$

$$75 \text{ g} - (37.5 \text{ g} + 11.2 \text{ g}) = 26.3 \text{ g}$$

It is improbable that the particle-size fractionation will yield a result that falls exactly between two textural classes. If this happens, there are complex rules for resolving which soil textural class applies. For your purposes, the simplest approach is to use the textural class that yields the finer texture (choose silty clay loam rather than clay loam, for example). Remember that the lines dividing textural classes are not absolutes, but approximations of where changes in soil texture can begin.

Example 5–4

If a mineral soil contains 80 percent sand, 15 percent silt, and 5 percent clay, what is its textural class?

Solution

In Figure 5–2, locate 80 percent on the axis corresponding to percent sand. Project a line toward the percent clay axis that is parallel to the opposing side (which corresponds to the percent silt axis). Locate 15 percent on the axis corresponding to percent silt. Project a line toward the percent sand axis that is parallel to the opposing side (which corresponds to the percent clay axis). The intersection of these two lines occurs within the textural class of this soil. For this particular distribution of particle sizes, Figure 5–2 illustrates that the textural class is designated as a loamy sand. To confirm your results, locate 5 percent on the percent clay axis and project a line parallel to the percent sand axis. This line should intersect the other two lines within the loamy sand region.

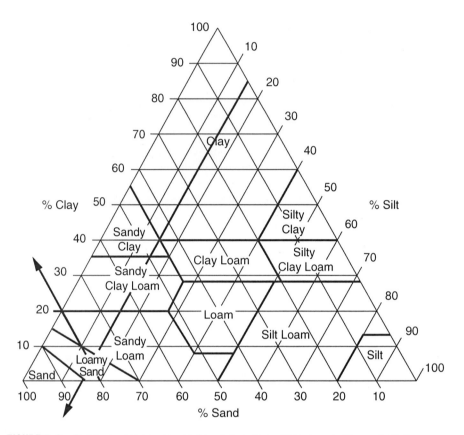

FIGURE 5–2 Soil textural triangle used to determine soil texture in Example 5-4.

Source. Soil survey staff. (1975). Soil taxonomy. In *Agricultural handbook no. 436. soil conservation service, USDA.* Washington, DC: U.S. Govt. Printing Office.

Reference

Thein, S. J., & Graveel, J. G. (1997). *Laboratory manual for soil science: Agricultural & environmental principles* (7th ed.). Dubuque, IA: W. C. Brown.

Sample Problems

Soil Texture—Particle Size and Distribution

1. What is the percent of sand, silt, and clay in a soil sample that contains 330 g of particles >0.05 mm in diameter, 330 g between 0.002 and 0.05 mm in diameter, and 340 g of particles <0.002 mm in diameter?

2. What is the percent of sand, silt, and clay in a soil sample that contains 500 g of particles >0.05 mm in diameter, 250 g between 0.002 and 0.05 mm in diameter, and 130 g of particles <0.002 mm in diameter?

3. What is the percent of sand, silt, and clay in a soil sample that contains 90 g of particles between 0.002 and 0.05 mm in diameter, 310 g of particles >0.05 mm in diameter, and 250 g of particles <0.002 mm in diameter?

4. If a 110 g soil sample contains 55 percent sand, 36 percent silt, and 9 percent clay in the mineral material <2.0 mm in diameter after sieving, what are the individual masses of sand, silt, and clay in this sample?

5. What is the percent of sand, silt, and clay in a soil sample that contains 700 g of particles >0.05 mm in diameter, 100 g between 0.002 and 0.05 mm in diameter, and 100 g of particles <0.002 mm in diameter?

6. If a 760 g soil sample contains 34 percent clay, 33 percent sand, and 33 percent silt, what is the mass of each individual fraction in this sample?

7. What is the percent of sand, silt, and clay in a soil sample that contains 140 g of particles >50 but <2000 μm in diameter, 38 g between 2 and 50 μm in diameter, and 22 g of particles <2 μm in diameter?

Soil Texture—Specific Surface Area

1. What is the total specific surface area of a 5 g soil sample that contains 60 percent sand, 30 percent silt, and 10 percent clay?

2. What is the total specific surface area of a 10 g soil sample that contains 15 percent sand, 60 percent silt, and 25 percent clay?

3. Which has more surface area, a parking lot 1 ha (10,000 m^2) in size or 100 g of soil (about a handful) that contains 35 percent clay?

Soil Texture—Using the Textural Triangle

1. What is the textural class of a soil that contains 33 percent sand, 33 percent silt, and 34 percent clay?

2. What is the textural class of a soil that contains 57 percent sand, 28 percent silt, and 15 percent clay?

3. What is the textural class of a soil that contains 10 percent sand, 68 percent silt, and 22 percent clay?

4. What is the textural class of a soil that contains 48 percent silt, 50 percent sand, and 2 percent clay?

5. What is the textural class of a soil that contains 40 percent clay, 52 percent sand, and 8 percent silt?

6. What is the maximum amount of silt a sandy loam can have?

7. What is the textural class of a soil that contains 20 percent sand and 60 percent clay?

8. If a soil with a texture of loam has 40 percent sand, what is the range of percent silt that is possible?

9. What is the minimum sand content of a loamy sand?

10. What is the textural class of any soil that contains 60 percent or more clay?

6

Bulk Density, Particle Density, and Porosity

OBJECTIVE

After completing this chapter you should be able to

- calculate bulk density and particle density.
- determine porosity if given bulk density and particle density values.

Overview

As does soil texture, soil structure has important influences on soil properties such as aeration and water content. The main effect of soil structure is on the development of porosity and porosity's role in conducting and storing water. Bulk density and porosity measurements can be easily made and can be very useful in developing meaningful interpretations of soil conditions for proper soil management. These measurements can also be used to determine soil water content. These exercises are designed to provide you with practice in determining soil bulk density and porosity from field and laboratory data.

Soil Density—Bulk Density and Particle Density

Two properties used to characterize the density of soil are **bulk density** (ρb) and **particle density** (ρp) (you may also see them abbreviated as BD and PD, or D_b

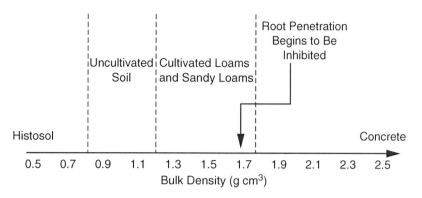

FIGURE 6–1 Typical bulk densities in soils and other materials (*adapted from Brady & Weil, 2002*).

and D_p, respectively). Bulk density and particle density are important measurements because they provide an indication of the compactness of a soil and its total pore space, which are vital structural characteristics for soil activity. A high bulk density (ρb) can indicate compacted or restricted layers in soil.

Bulk density is the dry mass of soil divided by the volume of soil. It reflects the density of the whole soil (soils and pores). The usual unit is g cm^{-3} or Mg m^{-3}. Bulk density ranges from <0.5 g cm^{-3} in organic soils (histosols) to >2.5 g cm^{-3} in materials like concrete. In typical cultivated soils, a bulk density >1.7 g cm^{-3} is indicative of compaction that can seriously retard root growth (Figure 6–1).

$$\text{Bulk Density } (\rho b) = \frac{\text{oven dry soil weight, g}}{\text{volume of soil solids and pores, cm}^3} \qquad \boxed{\text{6-1}}$$

There are several ways by which bulk density is determined. The most common is to pound a cylinder of known volume into soil, remove the cylinder, and dry the cylinder and soil in an oven at 105°C for 24 h to remove all the water. The weight of the cylinder + soil, less the weight of the cylinder, is the oven dry weight of soil. The volume of the soil is equal to the volume of the cylinder. Soil weight divided by soil volume yields the bulk density.

Example 6–1

What is the bulk density of a soil sample that weighs 1.50 g and occupies a volume of 0.75 cm^3?

Solution

$$\text{Bulk density} = \frac{1.50 \text{ g}}{0.75 \text{ cm}^3} = 2.0 \text{ g cm}^{-3}$$

Example 6–2

A cylinder, 7.5 cm in diameter and 7.5 cm in height, was used to collect an undisturbed soil sample. The container and soil weighed 505 g after the soil was dried, and the empty container weighed 75 g. What is the bulk density of this soil?

Solution

Step 1. Determine the mass of dry soil.

(Weight of the soil + container) − (weight of the container) = dry weight of soil

$$505 \text{ g} - 75 \text{ g} = 430 \text{ g}$$

Step 2. Determine the volume of soil.

This is equal to the volume of the cylinder. The formula for determining the volume of a cylinder is

$$V = h\pi r^2$$

where

V = volume (cm^3)
h = height (cm)
π = pi (approximately 3.142)
r = radius (equal to one-half the diameter of a circle)

Volume of the cylinder = (7.5 cm) π (7.5/2 cm)^2 = 331 cm^3

Step 3. Determine the bulk density.

Dry weight of soil ÷ volume of soil = bulk density

$$430 \text{ g} \div 331 \text{ cm}^3 = 1.30 \text{ g cm}^{-3}$$

Particle density (pp) is the dry mass of the soil's solid phase expressed on a volume basis. It omits the contribution of pore space to a soil's density and only reflects the solid material. The usual unit is also g cm^{-3} or Mg m^{-3}.

$$\text{Particle density } (pp) = \frac{\text{oven dry soil weight, g}}{\text{volume of soil solids, cm}^3} \qquad \boxed{\textbf{6-2}}$$

Particle density in most mineral soils varies from 2.60 to 2.70 g cm^{-3} because it reflects the average density of soil minerals. Since the dominant mineral in most soils is quartz (SiO_2), which has a density of 2.65 g cm^{-3}, for most purposes the particle density of soils is assumed to be 2.65 g cm^{-3} (Brady & Weil, 2002). Consequently, particle density has little practical significance except when it is used to calculate pore space.

Pore space, or **porosity**, is the proportion of the soil volume that is empty. You can calculate porosity if you know the bulk density (ρb) of a soil. Porosity is usually expressed as a percent.

$$\text{Porosity (\%)} = \left(1 - \frac{\rho b}{\rho p}\right)100 \qquad \textbf{6-3}$$

Another way of looking at this is that the percent solids in soil by volume is $100 \times$ (bulk density/particle density). The percent pore space by volume is $100 - \%\text{solid}$.

Example 6–3

What is the porosity of a soil core 10 cm high and 6 cm in diameter that weighs 500 g when dry?

Solution

The volume of the soil core is $(10 \text{ cm}) \, \pi \, (6/2 \text{ cm})^2 = 283 \text{ cm}^3$

The bulk density of the soil is $500 \text{ g}/283 \text{ cm}^3 = 1.77 \text{ g cm}^{-3}$

The particle density is assumed to be 2.65 g cm^{-3}

The porosity (%) $= \left[1 - \left(\dfrac{1.77 \text{ g cm}^{-3}}{2.65 \text{ g cm}^{-3}}\right)\right] \times 100 = 33\%$

Reference

Brady, N. C., & Weil, R. R. (2002). *The nature and property of soils*. Upper Saddle River, NJ: Prentice-Hall.

Sample Problems

Soil Density—Bulk and Particle Density

1. What is the bulk density of a soil with a dry mass of 2.60 g that occupies a volume 1.50 cm^3?
2. What is the bulk density of a dry soil sample with a mass of 30 g that completely occupies a cylinder 6 cm high and 4 cm in diameter?
3. Calculate the mass of a hectare furrow slice of soil that has a depth of 15 cm and a bulk density of 1.25 g cm^{-3}.
4. How many kilograms of soil with a bulk density of 1.10 g cm^{-3} are required to fill a box 1 m^2 with soil to a depth of 25 cm?

5. If you dug the equivalent of a 1 yd^3 in volume, how much soil would you end up moving if the bulk density was 1.8 g cm^{-3}?
6. What is the percent solids by volume in a soil sample that has a bulk density of 1.8 g cm^{-3} and a particle density of 2.65 g cm^{-3}?

Porosity

1. What is the percent pore space in a soil that has a bulk density of 1.6 g cm^{-3} and a particle density of 2.65 g cm^{-3}?
2. A soil core was taken to determine bulk density. The soil core had a volume of 80 cm^3 and a dry weight of 90 g. What is the percent pore space of this sample?
3. A soil sample taken from a compacted site and an uncompacted site were compared. The soil from the compacted site had a volume of 600 cm^3 and the weight of the soil and its container was 1300 g. The container weight was 300 g. The soil and container from the uncompacted site weighed only 900 g and the container weight was 180 g. The uncompacted soil had a volume 550 cm^3. Which soil has the greater porosity?
4. Given a porosity of 43 percent and a particle density of 3.1 g cm^{-3}, what is the bulk density?

7

Measurements
of Soil Water

OBJECTIVE

By the end of this chapter you should have learned how to

- calculate the water content in soil.
- determine the water status of soil at various descriptive levels.
- determine the direction of water flow based on water potential.

Overview

The pore space in soil is occupied by either air or water. You spent time learning how to calculate soil porosity primarily so that it would help you calculate how much water a soil could contain. Soil water influences virtually all soil processes, in addition to permitting plant growth. A soil's suitability for plant growth, in addition to its role as a rooting medium and supplier of growth nutrients, is largely determined by its capacity to collect, store, and release water (Thein & Graveel, 1997).

Measuring Soil Water Content

There are several ways that we can report soil water content—descriptively, by content, and by availability. We'll discuss methods of calculating available water in the next section. Descriptive terms for soil water content are saturation,

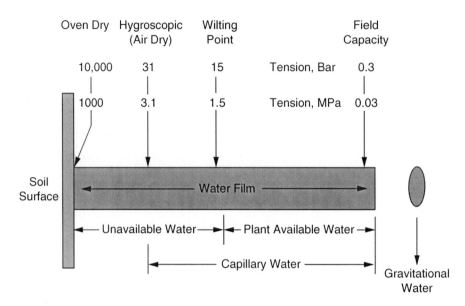

FIGURE 7–1   Types of water in soil (*adapted from Thein & Graveel, 1997*).

gravitational water, field capacity, plant available water, capillary water, plant unavailable water, and hygroscopic water.

Figure 7–1 shows a schematic diagram of how a water film surrounding the solids in soil might look. When the soil is saturated all the pores are filled with water and water drains through continuous pores to lower soil depths. Field capacity is the water content of soil when downward movement of gravitational water has ceased. In other words, it is the water content of soil when the largest soil pores have drained by the force of gravity. As with the other descriptive terms, it is usually reported as percent water by volume or weight. Available water is the water content in soil at field capacity minus the water content at the wilting point. The wilting point is the lowest end of the soil water content that is useful to plants. Capillary water is the water content at field capacity minus the water content in air-dry soil, or hygroscopic water content. Hygroscopic water is the water content in air-dry soil minus the water content in oven-dry soil.

We can put a value to each of these descriptive terms if we can calculate the percent water by weight or by volume. The percent water by weight (%wt) is called the **gravimetric water** (also θ_g). It is the weight of water in soil expressed as wet soil weight minus oven-dry soil weight divided by the oven-dry weight of soil:

$$\text{Percent soil water by weight (\%wt)} = \frac{\text{wet soil} - \text{oven-dry soil}}{\text{oven-dry soil}} \times 100 \quad \boxed{\textbf{7-1}}$$

Example 7–1

What is the gravimetric water content (%wt) of a moist soil sample weighing 80 g, which has a weight of 65 g after oven drying?

Solution

$$(\%\text{wt}) = \frac{80 \text{ g} - 65 \text{ g}}{65 \text{ g}} \times 100 = 23\%$$

Samples are usually oven dried in metal containers, so the weight of these containers must also be taken into account.

$$(\%\text{wt}) = \frac{(\text{wet soil} + \text{container}) - (\text{oven-dry soil} + \text{container})}{(\text{oven-dry soil} + \text{container})} \times 100 \quad \boxed{\textbf{7-2}}$$

Example 7–2

What is the gravimetric water content (%wt) of a moist soil sample in a metal container if, before oven drying, it weighs 105 g, after oven drying it weighs 95 g, and the metal container itself weighs 10 g?

Solution

$$\text{Percent soil water } (\%\text{wt}) = \frac{105 \text{ g} - 95 \text{ g}}{95 \text{ g} - 10 \text{ g}} \times 100 = 12\%$$

In many cases, experiments in soil science are conducted on field moist soils, and the results converted to an oven-dry soil basis after the fact. The equation to make this conversion is:

$$\text{Moist soil, g} \times \frac{100\%}{100\% + \text{percent soil water } (\%\text{wt})} = \text{oven-dry soil, g} \quad \boxed{\textbf{7-3}}$$

Example 7–3

If 100 g of field moist soil at 33 percent gravimetric water content was used in an experiment, how much oven-dry soil was actually used?

Solution

$$100 \text{ g} \times \frac{100\%}{100\% + 33\%} = 75 \text{ g oven-dry soil}$$

An easier approach to this calculation is to ignore the transformations into percent and simply divide the field moist weight by 1 + gravimetric water content (the simple ratio of water mass to soil mass).

Example 7–4

How much dry soil is in 75 g of a moist soil sample that has a gravimetric water content of 20 percent?

Solution

$$\text{Dry soil, g} = \frac{75 \text{ g moist soil}}{1.20} = 62.5 \text{ g oven-dry soil}$$

The reverse is also true. Once you know the gravimetric water content, you can add water to oven-dry soil to reach any moist water content you want.

Example 7–5

How much water would you have to add to 50 g of oven-dry soil to bring it to 33 percent gravimetric water content (a typical field moist water content)?

Solution

$$50 \text{ g oven-dry soil} \times 1.33 = 66.5 \text{ g moist soil at 33 percent} \\ \times \text{gravimetric water content}$$

$$66.5 \text{ g moist soil} - 50 \text{ g oven-dry soil} = 16.5 \text{ g water}$$

Add 16.5 g water to 50 g oven-dry soil to get a moist soil at 33 percent gravimetric water content.

You can double check your calculations in Example 7–5 by recalculating gravimetric water content

$$16.5 \text{ g water} \div 50 \text{ g oven-dry soil} = 0.33$$

$$0.33 \times 100 = 33\%$$

Once you know how to calculate percent water by weight, you can calculate the percent of any type of water characterized by the descriptive terms we previously used (Troeh & Thompson, 1993):

$$\text{Percent of any water type} = \frac{\text{weight 1} - \text{weight 2}}{\text{oven-dry weight}} \times 100 \qquad \boxed{\textbf{7-4}}$$

Example 7–6

If the weight of a soil is 30.0 g at field capacity, 27.2 g at the wilting point, 25.0 g when it is air dry, and 24.2 g when it is oven dry, what is the percent of water by weight for each soil?

Solution

$$\text{Percent water at field capacity} = \frac{30.0 \text{ g} - 24.2 \text{ g}}{24.2 \text{ g}} \times 100 = 24\%$$

$$\text{Percent water at wilting point} = \frac{27.2 \text{ g} - 24.2 \text{ g}}{24.2 \text{ g}} \times 100 = 12\%$$

$$\text{Percent available water} = \frac{30.0 \text{ g} - 27.2 \text{ g}}{24.2 \text{ g}} \times 100 = 12\%$$

If you know the bulk density, you can calculate the percent water by volume (%v):

$$\%v = \%wt \times \frac{\text{soil bulk density } (\rho b), \text{ g cm}^{-3}}{\text{density of water, g cm}^{-3}} \times 100 \qquad \textbf{7-5}$$

The density of water is usually taken to be 1.0 g cm^{-3}.

Example 7–7

If the bulk density of a soil is 1.50 g cm^{-3} and the gravimetric water content is 26 percent, what is the percent water by volume, and how much of the total pore space is filled by water?

Solution

$$\%v = \frac{\%wt \times (\rho b)}{(1.0 \text{ g cm}^{-3})}$$

$$\%v = \frac{26\% \times (1.50 \text{ g cm}^{-3})}{(1.0 \text{ g cm}^{-3})} = 39\%$$

$$\text{Total porosity} = \left[1 - \left(\frac{\rho b}{\rho p}\right)\right] \times 100$$

$$= \left[1 - \left(\frac{1.50 \text{ g cm}^{-3}}{2.65 \text{ g cm}^{-3}}\right)\right] \times 100 = 43\%$$

$$\text{Water-filled pore space} = \frac{\text{percent water by volume}}{\text{total porosity}} = \frac{39\%}{43\%} \times 100 = 91\%$$

Equivalent surface depth is a useful term that reflects the water content of a soil in terms of how deep a layer it would form if it were placed on the soil surface:

$$\text{Equivalent surface depth} = (\%v) \times (\text{soil thickness, cm})$$

$$(\text{cm of water per sample zone})$$

Example 7–8

How many centimeters of rain are required to wet a nearly dry soil 15 cm in depth to the equivalent of 36 percent water by volume?

Solution

$$0.36 \text{ cm water/cm soil} \times 15 \text{ cm depth} = 5.4 \text{ cm}$$

Measurements of Soil Water Availability

Two soils can have exactly the same gravimetric water content, yet one of the soils may have wilting plants and the other won't. Ultimately, it is not so much the water content that matters in soil biology as the water availability. Water potential (ψ) is a mathematical description that can be used to assess the availability of water in soil for plant and microbial growth. In Figure 7–1 you can see that for field capacity, wilting point, hygroscopic water, and oven-dry soil, an approximate water potential is given in units of either megapascals (MPa) or bars of tension. As the water potential decreases (there is greater tension), the availability of water decreases.

Many units are used to express water potential but megapascals is the preferred SI unit:

$$1 \text{ atmosphere (atm)} = 1.013 \text{ bar}$$
$$= 0.101 \text{ MPa}$$
$$= 1.033 \text{ kg cm}^{-2}$$
$$= 1033 \text{ cm of water}$$
$$= 76 \text{ cm of mercury}$$

Note that these are all units of pressure or suction.

Example 7–9

What is the equivalent water potential in atmospheres of water that is at 0.03, 1.5, and 3.1 MPa of tension?

Solution

There is 1 atm/0.101 MPa, so:

$$0.03 \text{ MPa} \times 1 \text{ atm/0.101 MPa} = 0.30 \text{ atm}$$
$$1.5 \text{ MPa} \times 1 \text{ atm/0.101 MPa} = 14.9 \text{ atm}$$
$$3.1 \text{ MPa} \times 1 \text{ atm/0.101 MPa} = 30.7 \text{ atm}$$

Water flows along free energy gradients, from high potential to low potential, so water potential is a measure of the potential energy of water in the soil relative to its potential energy in a pool of absolutely pure water (Papendick & Campbell, 1980). The lower the water potential, the lower the water's potential energy, and consequently its ability to flow.

Example 7–10

If there are two interconnected pools of water, one with a water potential of 1.5 atm and another with a water potential of -3.0 atm, in which direction will the water flow?

Solution

$$1.5 \text{ atm} > -3.0 \text{ atm}$$

Water will flow from the pool with a water potential of 1.5 atm to the pool with a water potential of -3.0 atm.

Water potential is composed of several components, which Figure 7–2 illustrates. Imagine that there is a pool of absolutely pure water in soil. Where will it flow? There is **capillary,** or **matric, potential** (ψm), which reflects the attraction of soil solids for water. It's the same effect you observe when you

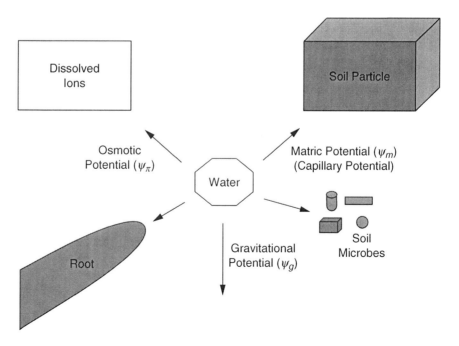

FIGURE 7–2 Sources of competition for water in soil (*adapted from Troeh & Thompson, 1993*).

watch a dry sponge soak up water from a tray of water. There is **gravitational potential** (ψg), which reflects the attraction of gravity for water (does water flow uphill or downhill?). When soils have drained to field capacity, gravitational potential is negligible. The **osmotic potential** ($\psi \pi$) reflects the attraction of dissolved solutes for water. In saturated soils, osmotic potential is not very important, but in soils that have a lot of salts, or have been recently fertilized, osmotic potential can have a big effect on the availability of water for plant roots and soil microorganisms. The **pressure potential** (ψp) reflects the effects of atmospheric pressure (or vacuum) as well as overlying water (hydrostatic pressure) on water movement. It is usually not a significant feature of soil water potential.

Roots and soil organisms have to compete for water with these attractive forces in soil. The water in living cells also has several components that can be used to describe its potential availability. Two are the same as soil—matric potential and osmotic potential. They reflect the tendency of cell water to be attracted to cell membranes and cell walls, and to be attracted to solutes dissolved in the cell cytoplasm. The matric potential in a living cell is usually considered to be a minor contributor to its overall water potential. Just like blowing a balloon up with air, filling a cell with water creates internal pressure as the water tries to escape its confined space. This is the pressure potential.

To summarize, water potential in soil is composed of three major components: matric potential, osmotic potential, and pressure potential (gravitational potential is ignored in most cases). Matric and osmotic potentials are usually given negative values because they make water less available. Pressure potential is given a positive value. The sum of these three components is the total water potential of a living cell or the soil:

$$\text{Total water potential } (\psi t) = \Sigma \text{matric potential} + \text{osmotic potential} + \text{pressure potential}$$

Example 7–11

What is the total water potential in a soil if the matric potential is -0.05 MPa, the osmotic potential is -1.0 MPa, and the pressure potential is negligible?

Solution

$$\text{Total water potential } (\psi t) = \Sigma \text{matric potential} + \text{osmotic potential} + \text{pressure potential}$$

$$\psi t = (-0.05 \text{ MPa}) + (-1.0 \text{ MPa}) = -1.05 \text{ MPa}$$

Soil organisms are in equilibrium with the water potential of their environment. If the water potential of their environment increases, the water potential of the organisms will increase. Likewise, if the water potential of their environment decreases, the water potential of the soil microbes will also decrease.

References

Papendick, R. I., & Campbell, G. S. (1980). Theory and measurement of water potential. *Water potential relations in soil microbiology.* (SSSA Special Publication Number 9, pp. 1–22). Madison, WI: Soil Science Society of America.

Thein, S. J., & Graveel, J. G. (1997). *Laboratory manual for soil science: Agricultural and environmental principles* (7th ed.). Dubuque, IA: W. C. Brown.

Troeh, F. R., & Thompson, L. M. (1993). *Soils and soil fertility* (5th ed.). New York: Oxford University.

Sample Problems

Measuring Soil Water Content

1. What is the gravimetric water content of a soil that weighs 9.8 g when wet and 7.8 g when oven dry?
2. How much oven-dry soil was used in an experiment if 8.0 g of soil at 87.5 percent gravimetric moisture content was initially weighed?
3. What is the volumetric content of a soil sample that has a bulk density of 1.4 g cm^{-3} and a gravimetric water content of 26 percent?
4. If a soil contains 33 percent water at field capacity, 20 percent water at the wilting point, and 5 percent water when it is air dry, what percent available water and what percent capillary water does it have?
5. For a 120 g moist soil sample in a drying tin that weighs 15 g, how much of the total pore space is filled with water if the final oven-dry weight is 100 g? Assume the bulk density is 1.4 g cm^{-3}.
6. What is the water-filled pore space of a soil that contains 33 percent water by volume and has a bulk density of 1.65 g cm^{-3}?
7. How much water is in a 1 ha furrow slice that contains 12 percent water by weight?
8. Find the percent air space in soil on a volume basis if bulk density = 1.35 g cm^{-3}, particle density = 2.65 g cm^{-3}, total porosity = 50%, and gravimetric water content = 25%.
9. Calculate the equivalent surface depth of water in soil that has a bulk density of 1.3 g cm^{-3} and has been irrigated to a depth of 20 cm if the percent gravimetric water in soil is 15%.
10. If the soil was dry initially, how many centimeters of rain would be required to fill 30 percent of the pore space in the top 15 cm of soil, assuming total porosity was 42 percent?

Measurements of Soil Water Availability

1. What is the equivalent water potential in bars for a soil with a water potential of -0.5 MPa?
2. Which has the greater water potential, a soil with a water potential of 0.3 MPa or one with a water potential of 0.01 MPa?
3. In which soil is water more available, one in which the water potential is -1.0 atm or one in which the water potential is -5.0 atm?
4. If you had to choose between two soils, one with 12 percent water by weight and a water potential of -1.5 MPa, and a soil with 10 percent water and a water potential of -1.0 MPa, which one would likely promote better plant growth?
5. What is the water potential of a soil in which $\psi m = -10$ MPa and $\psi \pi = -5$ MPa?
6. What is the water potential of a cell in which $\psi m = -0.005$ Pa, $\psi p = 20$ Pa, and $\psi \pi = -200$ Pa?
7. If a cell with a water potential of -2.0 MPa is placed in a soil with a water potential of -15 MPa, will the cell likely gain or lose water?
8. If a cell with a maximum pressure potential of 5 atm, a matric potential of -0.01 atm, and an overall water potential of -1.0 atm is placed in soil with a water potential of -10 atm, how much will the osmotic potential of the cell have to change if it is to come into equilibrium with the soil?

8

Water and Gas Transport

OBJECTIVE

Understanding how water flows through soil is critical to soil science. Properties such as infiltration rate, capillary flow, and gaseous diffusion determine whether ponding or runoff occur, how high a water table rises, and whether oxygen moves fast enough through soil to ensure that the soil stays well-aerated. By the end of this chapter, you'll be able to answer several questions.

- What is Darcy's law and how is it used?
- How high can water rise through capillary flow?
- How can you use Fick's law to describe gas transport through soil?

Water Flow in Soil

There are two types of water flow in soil—saturated flow and unsaturated flow. **Saturated flow** typically occurs after a heavy rainfall or irrigation. **Unsaturated flow** occurs when the larger pores in soil have drained and there is some airspace in soil. Saturated flow is described by **Darcy's law.** Unsaturated flow is frequently characterized by the **capillary rise equation.** We'll address each in turn.

Darcy's Law

One-dimensional water flow (vertically or horizontally) through saturated homogenous soil is calculated by using Darcy's law:

$$Q = \frac{KA\Delta H}{L}$$

8-1

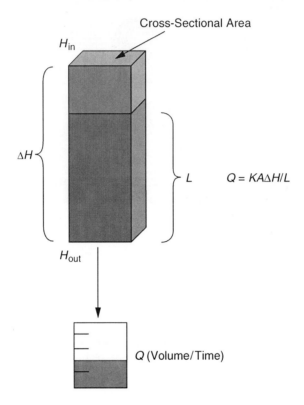

FIGURE 8–1 Diagrammatic representation of Darcy's law operating in a saturated soil system with a constant head of water.

where

Q = flow rate (cm^3 time^{-1})
K = saturated hydraulic conductivity of the soil (cm time^{-1})
A = cross-sectional area of flow (cm^2)
ΔH = head or potential causing flow (cm)
L = flow length (cm)

The easiest way to think about ΔH is in terms of the difference in height between where water enters soil and where water exits soil (Figure 8–1). The typical unit of K is cm s^{-1} or cm h^{-1}. Table 8–1 gives some typical K values for various saturated soils.

Example 8–1

A 10 cm column containing beach sand is irrigated with water. The cross-sectional area of the column is 25 cm^2, and the height of water above the column is maintained at 5 cm. If the K_{sat} of beach sand is 35.6 cm h^{-1}, what is the flux of water through this column?

TABLE 8–1 **Saturated Hydraulic Conductivity (K_{sats}) for Various Soils**

Soil Type	K_{sat} (cm s^{-1})	K_{sat} (cm h^{-1})
Beach sand	1×10^{-2}	36
Sandy soil	5×10^{-3}	18
Well-drained soil	5×10^{-4}	1.8
Poorly drained soil	5×10^{-5}	0.18
Impermeable soil	$<1 \times 10^{-8}$	$<3.6 \times 10^{-5}$

Note. Brady, N. C., & Weil, R. R. (2002). *The nature and properties of soil.* Upper Saddle River, NJ: Prentice-Hall.

Solution

It is easy to solve this problem by inserting the correct values into Darcy's law. ΔH is simply the height of the water plus the height of the sand column (10 cm + 5 cm):

$$Q = (35.6 \text{ cm h}^{-1}) (25 \text{ cm}^2) \frac{(15 \text{ cm})}{(10 \text{ cm})} = 1335 \text{ cm}^3 \text{ h}^{-1}$$

Example 8–2

To maintain a constant flux of 400 cm^3 h^{-1} through a soil column 15 cm high and 5 cm in diameter, how many centimeters of water would you have to maintain on the soil surface if the K_{sat} was 18 cm h^{-1}?

Solution

Step 1. First determine the cross-sectional area.

$$A = \pi r^2 = (\pi)\left(\frac{5}{2}\right)^2 = 19.6 \text{ cm}^2$$

Step 2. Substitute the known values into Darcy's law.

$$400 \text{ cm}^3 \text{ h}^{-1} = (18 \text{ cm h}^{-1}) (19.6 \text{ cm}^2) \frac{(? \text{ cm})}{(15 \text{ cm})}$$

Step 3. Rearrange the equation to solve for ΔH.

$$\frac{(15 \text{ cm})(400 \text{ cm}^3 \text{ h}^{-1})}{(18 \text{ cm h}^{-1})(19.6 \text{ cm}^2)} = 17 \text{ cm}$$

ΔH consists of the length of soil through which water flows (15 cm) plus the depth of the overlying water (2 cm), which is the value you want.

Capillary Rise

In unsaturated soil the soil pores are not completely filled with water, and usually all of the larger pores in soil have drained because of gravity. In other words, the soil is at field capacity or less. Water will nevertheless move by capillary flow. This is due to adhesive and cohesive properties of water. Capillary flow is described by the capillary rise equation, which relates the height to which water will rise above a water table in unsaturated soil to the diameter of pores in the soil environment:

$$h = \frac{2s \cos \alpha}{rDg}$$

8-2

where

h = capillary rise of water (m)
s = surface tension (0.0728 N m^{-1})
α = contact angle between water and a solid surface (°)
r = radius of the pore or capillary (m)
D = density of water (0.998 Mg m^{-3})
g = gravitational acceleration (9.81 J kg^{-1} m^{-1})

Fortunately, this equation can be simplified because most of the terms are constants. The working version of the capillary rise equation is as follows:

$$h \text{ (cm)} = \frac{0.15 \text{ cm}^2}{r \text{ (cm)}}$$

8-3

Example 8–3

How high will a column of water rise in a capillary with a diameter of 1 cm?

Solution

The radius of this capillary is 0.5 cm, therefore, using the capillary rise equation:

$$h \text{ (cm)} = \frac{0.15 \text{ cm}^2}{0.5 \text{ cm}} = 0.3 \text{ cm}$$

Example 8–4

How high will a column of water rise in a capillary with a diameter of 0.5 cm?

Solution

$$h \text{ (cm)} = \frac{0.15 \text{ cm}^2}{0.25 \text{ cm}} = 0.6 \text{ cm}$$

Notice the significance of the equation. Because the numerator is a constant, the capillary rise depends on the diameter of the capillary or pore; the smaller the diameter of the capillary, the greater the rise.

Gas Transport in Soil

Gas moves through soil by mass flow in response to concentration gradients. Because the concentration of O_2 in the atmosphere is always higher than in the soil, O_2 will flow into the soil as the O_2 is consumed. In contrast, CO_2 is almost always higher in the soil than the atmosphere, so it will flow out of the soil. Other gases, on the other hand, will flow into or out of the soil based on their concentration gradients, and usually this depends on whether the soil is producing or consuming these gases. The flow is most easily described by **Fick's law.**

Fick's Law

Fick's law describes gaseous diffusion through air and water-filled pores in soil:

$$J_g = D \frac{dc}{dx}$$

8-4

where

J_g = diffusive flux (mass flux across a unit area per unit time)
D = diffusion coefficient ($m^2 \ s^{-1}$)
dc/dx = concentration gradient of the gas (or the change in gas concentration over change in depth)

Table 8–2 shows some typical diffusion coefficients.

TABLE 8–2   Diffusion Coefficients in Air and Water

Gas	Diffusion Coefficient ($cm^2 \ s^{-1}$)
CO_2 in air	1.64×10^{-1}
CO_2 in water	1.6×10^{-5}
O_2 in air	1.98×10^{-1}
O_2 in water	1.9×10^{-5}
N_2 in water	2.3×10^{-5}

Note. Hillel, D. (1998). *Environmental soil physics.* San Diego, CA: Academic Press.

Example 8–5

What is the gaseous flux of O_2 through soil if the atmospheric concentration declines by 25 percent at a 50 cm depth? Assume the concentration of O_2 in the soil atmosphere is 0.3 kg m^{-3} and the diffusion coefficient is 1.98×10^{-5} m^2 s^{-1}.

Solution

The soil depth is 50 cm, the O_2 concentration at this depth is 0.75×0.3 kg m^{-3}, and the diffusion coefficient is given as 1.98×10^{-5} m^2 s^{-1}. Using Fick's law:

$$J_g = (1.98 \times 10^{-5} \text{ m}^2 \text{ s}^{-1}) \frac{(0.30 \text{ kg m}^{-3} - 0.225 \text{ kg m}^{-3})}{0.5 \text{ m}}$$

$$= 2.97 \times 10^{-6} \text{ kg m}^{-2} \text{ s}^{-1}$$

The soil environment is usually neither wholly wet nor wholly dry. So what is the best approximation for gas diffusion in soil itself? Although several formulas have been proposed, the easiest to use is **Penman's equation:**

$$\frac{D_s}{D_o} = 0.66 f_a \qquad \text{8-5}$$

where

D_s = apparent diffusion coefficient in soil
D_o = bulk air diffusion coefficient
f_a = air-filled porosity

Example 8–6

What is the gas flux of CO_2 from a 25 cm soil depth to the atmosphere, if the air-filled porosity is 25 percent, and the bulk air diffusion coefficient is 1.64×10^{-1} cm^2 s^{-1}? Assume the CO_2 concentration is 1 percent at 25 cm depth, and atmospheric CO_2 concentration is 360 ppm or 360 μL L^{-1}.

Solution

If there is 1 μmol/22.4 μL (a gas constant at standard temperature and pressure) and 1 μmol CO_2 has a molecular weight of 44 μg, then the bulk CO_2 concentration is 707 μg L^{-1} or 0.707 μg cm^{-3}. Likewise, the concentration of CO_2 in soil will be 19.6 μg cm^{-3}. The diffusion coefficient for soil will be:

$$D_s = (D_o) (0.66 f_a)$$

$$D_s = (1.64 \times 10^{-1} \text{ cm}^2 \text{ s}^{-1}) (0.66) (0.25) = 0.027 \text{ cm}^2 \text{ s}^{-1}$$

The concentration gradient $dc/dx = (19.6\ \mu g\ cm^{-3} - 0.707\ \mu g\ cm^{-3})/25\ cm = 0.76\ \mu g\ cm^{-4}$.

Solving for the rest of the equation gives:

$$J_g = (0.027\ cm^2\ s^{-1})\ (0.76\ \mu g\ cm^{-4}) = 0.02\ \mu g\ cm^{-2}\ s^{-1}$$

References

Brady, N. C., & Weil, R. R. (2002). *The nature and properties of soil* (p. 960). Upper Saddle River, NJ: Prentice-Hall.

Hillel, D. (1998). *Environmental soil physics* (p. 771). San Diego, CA: Academic Press.

Sample Problems

Darcy's Law

1. For a soil column 15 cm high containing very fine sandy soil with a K_{sat} of 1.8 cm h^{-1}, what is the flux of water if the soil is saturated and a constant water depth of 2 cm is maintained? Assume the cross-sectional area is 1 m^2.

2. If you had a soil core 10 cm in diameter and 100 cm long and the K_{sat} was 5×10^{-4} cm s^{-1}, what would be the saturated water flux through the core if you maintained a constant water level of 5 cm above the soil surface?

3. What would be the water flux through a horizontal soil core 25 cm long if it had a hydraulic conductivity of 0.07 cm h^{-1} and the reservoir delivering water to the core was 50 cm above it? Assume the core has a cross-sectional area of 10 cm^2.

4. What size sand filter would you need to supply an individual with 300 L of water per day if the required depth for adequate filtration was 30 cm and the K_{sat} of the sand was 36 cm h^{-1}? Assume you can maintain a hydraulic head of water 5 cm in depth.

Capillary Rise Equation

1. If the average pore diameter in soil above a water table was 50 μm, how high should the water rise (in cm) based on the capillary rise equation?

2. What is the maximum capillary rise of water in pores with diameters of 0.5 cm?

3. Show that water will rise higher in capillaries with diameters of 0.25 cm than 0.4 cm.

Fick's Law

1. What is the CO_2 flux through 50 cm of air if the atmospheric concentration of CO_2 is 350 ppm and a point source of CO_2 is 1000 ppm?
2. What is the diffusive flux of atmospheric O_2 through 25 cm of water if at the bottom of the column the O_2 concentration is 75 percent of the atmospheric concentration?
3. What is the O_2 flux through 10 cm of soil if the volumetric water content is 35 percent and the bulk density is 1.1 g cm^{-3}?
4. What is the minimum O_2 flux through 10 cm of soil if the O_2 concentration can decline by no more than 50 percent at this depth and the air-filled porosity is 50 percent? Assume the top of the soil is exposed to atmospheric concentrations of O_2.

9

Soil Temperature, Heat Capacity, and Conductivity

OBJECTIVE

In this chapter, you will learn how to

- determine the temperature rise in wet and dry soil.
- calculate heat capacity.
- evaluate thermal conductivity of soil.

Overview

Soil temperature has a profound effect on the rates of soil biological activity, mineral weathering, root growth, and evaporation. The capacity of soils to resist changes in soil temperature is very important and is referred to as the **heat capacity.** Heat, like water and gas, can also be transmitted through soil along gradients, in this case temperature gradients.

Describing Temperature in Soil

Three scales are commonly used to describe temperature in soil—centigrade (Celsius), Fahrenheit, and Kelvin. Of these, the most important scales used in soil science are centigrade and, to a lesser extent, Kelvin. Table 9–1 shows the relationship between the three scales in terms of some common benchmarks.

TABLE 9–1   Temperature Scales in Common Use for Soil Science

Scale	Absolute Zero	Freezing	Body Temperature	Boiling	Equations
Centigrade (°C)	−273	0	37	100	$°C = \frac{5}{9}(°F - 32)$
(Celsius)					$°C = K - 273$
Fahrenheit (°F)	−459	32	98.6	212	$°F = \frac{9}{5}°C + 32$
Kelvin (K)	0	273	310	373	$K = °C + 273$

Conversion between the different scales is relatively easy. The temperature in Kelvin is simply the temperature in °C + 273.

Example 9–1

What temperature in Kelvin is 60°C?

Solution

$$60 + 273 = 333 \text{ K}$$

Conversions between Celsius and Fahrenheit are routinely incorporated into most programmable calculators, but there are simple formulas to use as well.

Example 9–2

Convert 75°F to Celsius.

Solution

$$\frac{5}{9}(°F - 32) = °C$$

$$\frac{5}{9}(75 - 32) = 24\,°C$$

Example 9–3

If the temperature is 4°C, should you be wearing a sweater or shorts based on the temperature in Fahrenheit?

Solution

$$\frac{9}{5}°C + 32 = °F$$

$$\left(\frac{9}{5}\right)(4) + 32 = 39\,°F$$

A little chilly for shorts.

Heat Capacity

Heat capacity is the amount of energy (calories or joules) that an object must absorb or lose for its temperature to change by 1°. It is generally expressed on a weight basis as cal g^{-1} or J g^{-1}, or on a volume basis. The volumetric heat capacity of a soil is the change in heat content of a unit volume of soil per unit change in temperature expressed as cal cm^{-3} K^{-1} or $°C^{-1}$.

$$C_{total} = \Sigma f_{soil}C_{soil} + f_{water}C_{water} + f_{air}C_{air} \qquad \textbf{9-1}$$

where

  C = heat capacity of each constituent
  f = volumetric fraction of each constituent

You can see in Table 9–2 that the contribution of air to the heat capacity of soil is minimal. So Equation 9-1 can be adjusted so that it only reflects the major contributors to heat capacity in soil—minerals (m), organic matter (om), and water (w).

$$C = f_{minerals}C_{minerals} + f_{om}C_{om} + f_{water}C_{water} \qquad \textbf{9-2}$$

$$C = f_m 0.48 + f_{om}0.60 + f_w 1.0 \qquad \textbf{9-3}$$

Equation 9–3 reflects typical values for heat capacity of the different soil constituents from Table 9–2.

Example 9–4

What is the heat capacity of a soil sample that contains 10 g of minerals, 1 g of organic matter, and 2 g of water? Use 2.65 g cm^{-3} to estimate the density of minerals in soil (i.e., the particle density). Use 1.3 g cm^{-3} to estimate the density of organic matter. Use 1 g cm^{-3} to estimate the density of water.

TABLE 9–2 Volumetric Heat Capacity (C) and Thermal Conductivity (K) of Soil Constituents

Constituent	Density (ρ) (g cm^{-3})	Heat Capacity (C) (cal cm^{-3} K^{-1})	Thermal Conductivity (K) (cal cm^{-1} s^{-1} K^{-1})
Quartz	2.66	0.48	0.021
Basalt, slate	2.4–2.5	0.48	0.004–0.008
Chalk	1.8–2.0	0.3–0.4	0.0015–0.003
Other minerals	2.65	0.48	0.007
Sand	1.5	0.30	0.0045–0.0055
Organic matter	1.30	0.60	0.0006
Water	1.00	1.00	0.00137
Air	0.00125	0.003	0.00006

Solution

Because our equation for heat capacity contains units of volume, convert the mass values to volume values using the average density of each component.

$$\frac{10 \text{ g minerals} \times 1 \text{ cm}^3}{2.65 \text{ g}} = 3.77 \text{ cm}^3$$

$$\frac{1 \text{ g organic matter} \times 1 \text{ cm}^3}{1.3 \text{ g}} = 0.77 \text{ cm}^3$$

$$\frac{2 \text{ g water} \times 1 \text{ cm}^3}{1 \text{ g}} = 2 \text{ cm}^3$$

$$
\begin{aligned}
\text{Heat capacity } (C) &= (3.77 \text{ cm}^3)\,(0.48 \text{ cal cm}^{-3}\,°\text{C}^{-1}) \\
&\quad + (0.77 \text{ cm}^3)\,(0.60 \text{ cal cm}^{-3}\,°\text{C}^{-1}) \\
&\quad + (2 \text{ cm}^3)\,(1.0 \text{ cal cm}^{-3}\,°\text{C}^{-1}) \\
&= 4.27 \text{ cal } °\text{C}^{-1}
\end{aligned}
$$

One of the important things to recognize about heat capacity is that it is strongly affected by water. The more water you have in a soil, the more energy it will take to raise the soil temperature. That is one reason why wetter soils tend to be cooler than drier soils.

Another useful point to recognize is that between 0 and 5 percent organic matter, which are typical values for many soils, the heat capacity of soil solids can be approximated as 0.2 cal g^{-1} °C. So, heat capacity can be estimated if you can determine the solids content based on bulk density and the water content based on gravimetric or volumetric water content.

Thermal Conductivity

Radiation, convection, and conduction are the principle modes of energy transfer in soil. Radiation is the emission of energy by electromagnetic waves from all bodies warmer than 0 K. Convection is movement of a heat carrying mass such as a current or wind. Conduction is the propagation of heat within a body by internal molecular motion, i.e., transfer of heat by kinetic energy. To use an analogy, you start a fire by conduction when you rub two sticks together, feel its radiant energy when you stand next to it, back away when the wind blows the fire in your direction, and burn your hand by conduction again if you are unfortunate enough to pick up an iron poker from the fire without gloves.

Heat moves through the soil mainly by conduction. Conductive heat flow depends on the temperature gradient in soil. It is described by Fourier's heat flow equation:

$$Q = (A)\,(K)\left(\frac{\Delta T}{\Delta x}\right)$$

9-4

where

Q = thermal flux, energy conducted across a unit area per unit time, which has units of cal or J cm^{-2} s^{-1}

A = cross-sectional area, usually in cm^2

K = thermal conductivity

$\Delta T/\Delta x$ = change in temperature with change in distance, that is, the temperature gradient

In the case of soil, $\Delta T/\Delta x$ is denoted as the change in temperature (°C) with change in depth (cm).

Thermal conductivity (K) is a measure of a material's ability to transmit heat. Metals are excellent heat conductors, which is one reason why during a fire you should feel the doorknob before opening a door; it will reflect the temperature in the other room. Air is a poor conductor, which is why double pane glass has better insulation capacity than single pane glass, and why you can stick your hand in an oven at 450°F without ill effect but wouldn't dare do so in water at only 212°F. The thermal conductivity of various materials is shown in Table 9–2.

Example 9–5

What is the thermal flux through a uniform layer of sand 15 cm thick if the temperature at the top of the sand is 37°C, the temperature at the bottom of the sand is 25°C, and the surface area of the sand is 10 cm^2?

Solution

$$Q = (A)\,(K)\left(\frac{\Delta T}{\Delta x}\right)$$

$$Q = (10\ \text{cm}^2)\,(0.0045\ \text{cal cm}^{-1}\,\text{s}^{-1}\,{}^\circ\text{C}^{-1})\left(\frac{37 - 25\,{}^\circ\text{C}}{15\ \text{cm}}\right)$$

$$Q = 0.036\ \text{cal s}^{-1}$$

Note that in this example we have reported the temperature units in Celsius rather than Kelvin as there is no difference in the outcome of the calculation by this change in scales.

Thermal conductivity and temperature gradients have many applications in determining the depth to which soil freezes and thaws and the temperature changes that occur in soil at various depths on a daily basis. Figure 9–1 illustrates how the maximum and minimum daily temperature in soil differs markedly from that of air during a season. However, soil is a mixed medium, so calculating the thermal conductivity is much more complicated than the case illustrated here. So, these topics are beyond the scope of this book.

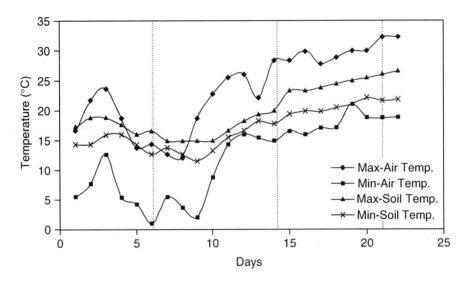

FIGURE 9–1 One of the practical significances of temperature fluctuation in soil is that temperature variability in soil is less than in air, as this figure illustrates. The dashed lines represent periods of rain [adapted from Gandhapudi, S. (2004). Managing fecal bacteria and nutrient contamination in poultry manure-amended sod by mowing and alum addition. M.S. Thesis. Lexington, KY: University of Kentucky.]

Sample Problems

Temperature Conversions

1. How many degrees Celsius is 298 K?
2. If the thermometer reads 45°F and you need to report it in degrees Celsius, what temperature do you write?
3. −50°C is what temperature in degrees Fahrenheit?
4. Convert 50°F to Kelvin.

Heat Capacity

1. What is the heat capacity of a soil sample that contains 7 g of minerals, 2 g of organic matter, and 1.5 g of water?
2. If an oven-dry soil sample with 1.5 percent organic matter has a volume of 10 cm³ and a bulk density of 1.1 g cm³, what is its heat capacity?
3. If 15 cm³ of soil weighing 20 g has a gravimetric water content of 25 percent and an organic matter content of 2.0 percent, what is its heat capacity?

4. By how much should the temperature of a soil sample rise if it is exposed to 0.5 cal of energy and the sample consists of 20 g minerals, 1.5 g organic matter, and 2 g of water?

Thermal Conductivity

1. What is the thermal flux through a column of water 10 cm high if it has a temperature gradient of 15°C from top to bottom? Assume the surface area is 20 cm².
2. What is the thermal flux through a piece of chalk 5 cm tall if one end is sitting on a hot plate at 100°C and the other is exposed to room temperature (25°C)? Assume the surface area is 1 cm².
3. If the thermal conductivity of quartz is 0.021 cal cm^{-1} s^{-1} K^{-1} and the thermal conductivity of organic matter is 0.0006 cal cm^{-1} s^{-1} K^{-1}, how tall must a column of each material be to maintain a constant thermal flux of 0.005 cal cm^{-2} s^{-1} if the temperature gradient is 10°C?
4. If you wanted to insulate your earth-buried home to maintain a constant temperature difference between the inside and outside environments at 30°C and a thermal flux of no more than 1.0 cal cm^{-2} s^{-1}, how much more sand would you have to use than organic matter?

Section III

Problem Solving in Soil Biochemistry

10

pH, Buffers, and Buffering

OBJECTIVE

In this chapter you will learn

- definitions of acids and bases.
- pH calculations.
- differences between normality and molarity.
- how to neutralize strong acids and bases.
- calculations for titrating weak acids and bases.
- how to create buffers of varying strength.

Overview

Buffers are critical in soil science because so much of the chemical and biological response of the soil environment is influenced by soil pH and its fluctuations. Buffers do exactly what their name implies. They buffer, or protect, the environment from sudden change. In this chapter you will look at some calculations you'll need to know to work with pH and buffers in soil science.

Acids, Bases, and Water

The Brønsted definition of an acid is a substance that donates protons (H^+) and a base as a substance that accepts H^+.

$$HA \quad + \quad B^- \quad \leftrightarrow \quad A^- \quad + \quad HB$$

$$\text{Acid}_1 \qquad \text{Base}_2 \qquad \text{Base}_1 \qquad \text{Acid}_2$$

10-1

A strong acid completely ionizes to yield the corresponding acid and base pairs. For example, HCl is a strong acid because in water:

$$HCl \quad + \quad H_2O \quad \leftrightarrow \quad Cl^- \quad + \quad H_3O^+$$

$$\text{Acid}_1 \qquad \text{Base}_2 \qquad \text{Base}_1 \qquad \text{Acid}_2$$

10-2

The hydronium ion (H_3O^+) and H^+ are identical for all intents and purposes. Likewise, a strong base is a compound that also ionizes completely in water:

$$NaOH \rightarrow Na^+ + OH^-$$

10-3

Water can dissociate (come apart) to form H^+ and OH^-. In pure water, the H^+ concentration is 10^{-7} M. For every mole of H^+ produced, 1 mol of OH^- is produced, and thus the OH^- concentration is 10^{-7} M. Remember that 1 mol of any substance is 6.023×10^{23} molecules of that substance, and that the concentration of most substances in water is quantified in terms of molarity (M) or moles per liter (mol L^{-1}).

pH

The shorthand notation for describing how much H^+ is in water is called **pH.** pH stands for the negative logarithm of the hydrogen-ion concentration (abbreviated as [H^+]) and is expressed as:

$$pH = -\log[H^+]$$

10-4

pOH stands for the negative logarithm of the hydroxyl concentration:

$$pOH = -\log[OH^-]$$

10-5

Because pH + pOH = 14, if you know pH, you also know pOH. The important point to remember is that the concentration of H^+ increases as the pH declines, and decreases as the pH rises.

Example 10–1

What is the pH of a 0.002 M solution of HCl?

Solution

HCl is a strong acid that completely dissociates. So, the [H^+] will be nearly 0.002 M.

$$pH = -\log[0.002 \text{ M}] = 2.7$$

Example 10–2

What is the pH of a 10^{-8} M solution of HCl?

Solution

$-\log[10^{-8} \text{ M}] = 8$, but the pH $\neq$ 8. Remember that the ionization of pure water gives a [H^+] of 10^{-7} M. So, the pH would be $-\log (10^{-7} + 10^{-8}) = 6.96$

Molarity and Normality

You have been dealing with problems that look at the concentration of H^+ in terms of its **molarity** (M = moles/liter). An important concept in buffering and acid/base chemistry is **normality** (N = equivalents/liter). Just as moles = grams/molecular weight, equivalents = grams/equivalent weight. Equivalent weight and molecular weight are related by the equation:

$$EW = \frac{MW}{n}$$

10-6

where

    EW = equivalent weight
    MW = molecular weight
       n = number of H^+ or OH^- that can be replaced if you're dealing with acids and bases, or number of electrons gained or lost per molecule if you're dealing with oxidation and reduction reactions or molecular charge

Another way of expressing this is as:

$$\text{Equivalents} = n \times \text{moles}$$

Molarity and normality are related by the equation:

$$N = nM$$

10-7

Example 10–3

How many moles or equivalents are in 28 g NaOH?

Solution

The molecular weight of NaOH is 40 g mol^{-1}.

    The number of moles = 28 g/40 g mol^{-1} = 0.7 mol
    The equivalent weight of NaOH = molecular weight/n
    Only one OH^- is present in NaOH, so $n = 1$
    Therefore, there are also 0.7 eq in 28 g of NaOH

Example 10–4

What is the normality of a 0.03 M solution of H_2SO_4?

Solution

There are two ionizable H^+ associated with H_2SO_4, so $n = 2$.

$$N = nM, \text{ so } N = (2)(0.03 \text{ M})$$

$$N = 0.06$$

Example 10–5

How many moles and equivalents of Ca^{2+} are in 50 g of $CaCl_2$?

Solution

There are 111 g $CaCl_2$/mol $CaCl_2$.

$$(50 \text{ g CaCl}_2) \times \frac{\text{mol CaCl}_2}{111 \text{ g CaCl}_2} = 0.45 \text{ mol CaCl}_2$$

There is one atom of Ca in $CaCl_2$ so there are also 0.45 mol Ca. Equivalents $= n \times$ moles. Since the charge on Ca $= 2+$, $n = 2$. So, 0.45 mol Ca $\times 2 = 0.90$ eq Ca.

Example 10–6

What is the normality of a 0.05 M solution of $CaCl_2$?

Solution

The charge on Ca is 2+, so $n = 2$.

$$N = nM, \text{ so } N = (2)(0.05 \text{ M})$$
$$N = 0.10 \text{ CaCl}_2$$

Neutralizing Strong Acids and Bases

One of the major procedures for assessing CO_2 evolution from soil is to trap the CO_2 in a strong base solution such as 1 M NaOH and subsequently titrating the solution with a strong acid such as HCl. Titration stops when all the remaining OH^- has been neutralized. This also becomes important in considering liming reactions to raise soil pH. So, it is important to understand neutralization reactions quantitatively.

The total equivalents of acid required to neutralize the total equivalents of base are exactly equal to one another. You can relate the two by a simple equation:

$$(N_{acid})(\text{Volume}_{acid}) = (N_{base})(\text{Volume}_{base}) \tag{10-8}$$

Example 10–7

How many milliliters of 0.3 M H_2SO_4 are required to exactly neutralize 250 mL of 0.75 M KOH?

Solution

The first thing to do in working neutralization reactions is to put everything on the same equivalent basis. 0.75 M KOH = 0.75 N KOH because normality and molarity are identical for compounds that dissociate to yield ions with a single charge. On the other hand, the normality of 0.3 M H_2SO_4 is 0.6 because $N = nM$ and H_2SO_4 has two protons to exchange. To answer the question, you write the equation:

$$(N\ H_2SO_4)(\text{Volume } H_2SO_4) = (N\ KOH)(\text{Volume } KOH)$$

$$(0.6\ N\ H_2SO_4)(\text{mL } H_2SO_4) = (0.75\ N\ KOH)(250\ \text{mL } KOH) = 312.5\ \text{mL } H_2SO_4$$

Titrating Weak Acids—The Henderson-Hasselbach Equation

Strong acids and bases completely dissociate when they are added to water, but weak acids like acetic acid (CH_3COOH), for example, do not (Segal, 1976). Weak acids form equilibrium relationships in water:

$$HA \leftrightarrow H^+ + A^- \hspace{2cm} \textbf{10-9}$$

where HA is the undissociated acid, and H^+ and A^- represent the acid and base components of the acid, respectively.

The extent to which the weak acid dissociates is described in Equation 10–10:

$$K_a = \frac{[H^+][A^-]}{[HA]} \hspace{2cm} \textbf{10-10}$$

where K_a is the dissociation constant of the weak acid and is usually a very small number. Like pH, it is frequently reported as pK_a (the negative logarithm of the dissociation constant K_a). The pH of a weak acid is described by the equation:

$$pH = \frac{(pK_a + pHA)}{2} \hspace{2cm} \textbf{10-11}$$

Example 10–8

What is the pH of a 0.15 M solution of a weak acid (HA) that has a K_a of 10^{-5} M?

Solution

The pK_a of 10^{-5} M = 5

$$pHA = -\log[0.15\ M] = 0.82$$

$$pH = \frac{(pK_a + pHA)}{2}$$

$$pH = \frac{(5 + 0.082)}{2} = 2.54$$

Example 10–9

What is the pH of a 10 mM solution of NH_4^+ (a weak acid), which has a pK_a of 9.2?

Solution

$$pHA = -\log (0.01 \text{ M NH}_4^+) = 2$$

$$pH = \frac{(pK_a + pHA)}{2}$$

$$pH = \frac{(9.2 + 2.0)}{2} = 5.6$$

Values for pK_a of various buffering systems are listed in Appendix 1.

For a weak acid, the pH at any time can be described by the **Henderson-Hasselbach equation:**

$$pH = pK_a + \log \frac{[A^-]}{[HA]} \qquad \text{\textbf{10-12}}$$

When the weak acid solution is exactly half titrated, so that [HA] = [A⁻], the pH = pK_a.

Example 10–10

What is the pH of 400 mL of weak acid (0.15 M) titrated with 0.2 M KOH, after 50 mL of KOH has been added? Assume the pK_a of the acid is 5.5.

Solution

$$50 \text{ mL} = 0.05 \text{ L KOH}$$

$$(0.05 \text{ L KOH}) \times (0.2 \text{ M KOH}) = 0.01 \text{ mol KOH} = 0.01 \text{ eq KOH}$$

$$(0.15 \text{ M HA}) \times (0.4 \text{ L HA}) = 0.06 \text{ mol HA} = 0.06 \text{ eq HA}$$

$$0.06 \text{ mol} - 0.01 \text{ mol} = 0.05 \text{ mol HA left}$$

$$pH = pK_a + \log \frac{[A^-]}{[HA]}$$

$$pH = 5.5 + \log \frac{(0.01 \text{ mol A}^-)}{(0.05 \text{ mol HA})} = 4.8$$

Buffers and Buffering

Common buffers are mixtures of a conjugate acid and base (CH_3COOH and CH_3COO^-, for example). Mixtures of these weak acids or bases resist pH change by accepting H^+ or OH^- added to the system. If H^+ is added, for example, CH_3COO^- will convert to CH_3COOH, while if OH^- is added, CH_3COOH will further dissociate to release H^+ and subsequently form H_2O.

pK_a is important in buffered systems (hence, the reason you have looked at it in detail) because the ability of a buffer to resist changes in pH is greatest around the pK_a. If you want a buffer that will strongly resist pH changes and keep a solution of approximately pH 7, you choose a buffer with a pK_a of approximately 7 rather than one with a pK_a of 5.5.

Example 10–11

How much CH_3COOH and CH_3COO^- are in a 0.25 M buffer that has a pH of 6.0 given that the K_a for acetate buffers is 1.7×10^{-5}?

Solution

$$pK_a = -\log [1.7 \times 10^{-5}] = 4.77$$

$$pH = pK_a + \log \frac{[A^-]}{[HA]}$$

$$6 = 4.77 + \log \left(\frac{[CH_3COO^-]}{[CH_3COOH]} \right)$$

If the concentration of the buffer is 0.25 M, and the total concentration of CH_3COOH and $CH_3COO^- = 0.25$ M, you can then solve for the concentration of either CH_3COO^- or CH_3COOH algebraically.

$$0.25 \text{ M} = [CH_3COOH] + [CH_3COO^-]$$

$$0.25 \text{ M} - [CH_3COOH] = [CH_3COO^-]$$

$$6 = 4.77 + \log \left(\frac{0.25 \text{ M} - [CH_3COOH]}{[CH_3COOH]} \right)$$

$$1.23 = \log \left(\frac{0.25 \text{ M} - [CH_3COOH]}{[CH_3COOH]} \right)$$

$$10^{1.23} = \left(\frac{0.25 \text{ M} - [CH_3COOH]}{[CH_3COOH]} \right)$$

$$16.98 = \left(\frac{0.25 \text{ M} - [CH_3COOH]}{[CH_3COOH]} \right)$$

$$16.98 [CH_3COOH] = (0.25 \text{ M} - [CH_3COOH])$$

$$16.98 [CH_3COOH] + [CH_3COOH] = 0.25 \text{ M}$$

$$17.98 [CH_3COOH] = 0.25 \text{ M}$$

$$[CH_3COOH] = \frac{0.25 \text{ M}}{17.98} = 0.014 \text{ M}$$

If $[CH_3COOH] = 0.014$ M, then $[CH_3COO^-] = 0.236$ M

$$6 = 4.77 + \log \frac{0.236}{0.014}$$

Most effective environmental buffers have a pK_a ranging from 3 to 9. Preparing buffers for laboratory use and chemical analysis is an important activity in soil science.

Example 10–12

How do you prepare 2 L of 0.3 M acetate buffer at pH 5.0 if your starting ingredients are sodium acetate trihydrate ($NaCH_3COO^- \cdot 3H_2O$, MW = 136) and 2 M acetic acid (CH_3COOH)?

Solution

$$pH = pK_a + \log \frac{[A^-]}{[HA]}$$

$$5.0 = 4.7 + \log \frac{[A^-]}{[HA]}$$

$$A^- + HA = 0.3 \text{ M}$$

$$A^- = 0.3 \text{ M} - HA$$

$$0.3 = \log \frac{[0.3 \text{ M} - HA]}{[HA]}$$

$$1.995 \text{ HA} = 0.3 \text{ M} - HA$$

$$2.995 \text{ HA} = 0.3 \text{ M}$$

$$HA = 0.10 \text{ M}$$

$$A^- = 0.20 \text{ M}$$

The HA comes from the acetic acid. To calculate how much you need:

$$(2 \text{ L}) \times (0.10 \text{ mol L}^{-1} \text{ HA}) = 0.20 \text{ mol HA}$$

0.20 mol HA/2 mol L^{-1} acetic acid = 0.10 L of acetic acid.
The A^- comes from the sodium acetate trihydrate, which has a molecular weight of 136 g mol^{-1}.

$$(0.20 \text{ mol L}^{-1} A^-) \times 2 \text{ L} = 0.40 \text{ mol } A^- \times 136 \text{ g mol}$$
$$= 54.4 \text{ g sodium acetate trihydrate}$$

So, to make the buffer, dissolve 54.4 g of sodium acetate trihydrate in some water, add 100 mL of 2 M acetic acid, and dilute to 2 L, and write down the amounts you need so you don't have to go through all the calculations the next time you make the buffer.

An easier approach to buffer preparation is to realize that when the acid and base conjugates of a buffer system are in equimolar concentrations, $pH = pK_a$. If you want to raise the pH by one unit from the pK_a, there should be

10 times as much of the acid form as the base form. A useful observation is that if you make a dilute solution with only the acid form of the compound, the pH will be approximately two units less than the pK_a. Conversely, if you make a dilute solution with only the base form of a compound, the pK_a will be approximately two units greater than the pK_a.

Example 10–13

Prepare 1 L of a 0.2 M phosphate buffer at pH 7.2.

Solution

The $H_2PO_4^- \leftrightarrow H^+ + HPO_4^{2-}$ buffer system has a pK_a of 7.2. To obtain a pH 7.2 buffer, you simply have to have equimolar amounts of $H_2PO_4^-$ and HPO_4^{2-} with a total molar concentration of 0.2 M. In other words $[H_2PO_4^-] = [HPO_4^{2-}] = 0.1$ M. Potassium phosphate monobasic (KH_2PO_4, MW = 136) and potassium phosphate dibasic (K_2HPO_4, MW = 174) are typically used to provide the $H_2PO_4^-$ and HPO_4^{2-}, respectively.

If the molar concentration of each form is 0.1 M and 1 L of buffer is required, then 0.1 mol of each chemical is used:

$$KH_2PO_4 = 136 \text{ g mol}^{-1} \times 0.1 \text{ mol} = 13.6 \text{ g}$$
$$K_2HPO_4 = 174 \text{ g mol}^{-1} \times 0.1 \text{ mol} = 17.4 \text{ g}$$

So, to make this buffer, dissolve 13.6 g of KH_2PO_4 and 17.4 g of K_2HPO_4 in water and dilute to 1 L.

An alternative method is to make equimolar stock solutions of the acid and base conjugates of a buffering system and then add appropriate volumes to achieve the specific pH values at the required concentration.

Example 10–14

Prepare 1 L of 0.02 M phosphate buffer, pH 6.8 from solutions of 0.2 M KH_2PO_4 and K_2HPO_4.

Solution

The stock concentrations are 10 times more concentrated than the final buffer concentration, so you can make 100 mL of the stock at the appropriate pH and dilute it 10-fold with little effect on pH.

$$pH = pK_a + \log \frac{[A^-]}{[HA]}$$

$$6.8 = 7.2 + \log \frac{[HPO_4^{2-}]}{[H_2PO_4^-]}$$

$$-0.4 = \log \frac{[HPO_4^{2-}]}{[H_2PO_4^-]}$$

The total volume of stock solutions used will be 100 mL. Simply solve the ratio for either the acid or base form on a volume basis.

$$\text{mL HPO}_4^{2-} + \text{mL H}_2\text{PO}_4^{-} = 100 \text{ mL}$$

$$\text{mL HPO}_4^{2-} = 100 \text{ mL} - \text{mL H}_2\text{PO}_4^{-}$$

$$-0.4 = \log \frac{(100 \text{ mL} - \text{mL H}_2\text{PO}_4^{-})}{(\text{mL H}_2\text{PO}_4^{-})}$$

$$10^{-0.4} = 0.398 = \frac{(100 \text{ mL} - \text{mL H}_2\text{PO}_4^{-})}{(\text{mL H}_2\text{PO}_4^{-})}$$

$$0.398 \,(\text{mL H}_2\text{PO}_4^{-}) = (100 \text{ mL} - \text{mL H}_2\text{PO}_4^{-})$$

$$1.398 \,(\text{mL H}_2\text{PO}_4^{-}) = 100 \text{ mL}$$

$$\text{mL H}_2\text{PO}_4^{-} = 71.54 \text{ mL}$$

$$\text{mL HPO}_4^{2-} = 28.5 \text{ mL}$$

To make this buffer, add 28.5 mL of 0.2 M K_2HPO_4 to 71.5 mL of 0.2 M KH_2PO_4 and dilute to 1 L.

An advantage of using stock solutions to prepare buffers is that once you've calculated the appropriate volumes of acid and base stock for a particular pH, you should never have to calculate them again. Appendix 2 has recipes for preparing various buffers spanning the range of pH typical in soil environments using this approach.

A third alternative for buffer preparation is to use the appropriate concentration of acid or base for a buffering system and titrate with a strong acid or base (HCl or NaOH, for example) until the appropriate pH is obtained.

Example 10–15

Prepare 1 L of 0.1 M phosphate buffer, pH 7.3, starting with KH_2PO_4 and a solution of 1 M NaOH.

Solution

The molecular weight of KH_2PO_4 is 136 g mol. One liter of a 0.1 M buffer therefore requires:

$$(1 \text{ L}) \times (0.1 \text{ mol KH}_2\text{PO}_4 \text{ L}^{-1}) = 0.1 \text{ mol KH}_2\text{PO}_4$$

$$(0.1 \text{ mol KH}_2\text{PO}_4) \times (136 \text{ g KH}_2\text{PO}_4 \text{ mol}^{-1}) = 13.6 \text{ g KH}_2\text{PO}_4$$

All of the phosphate is supplied by KH_2PO_4. The pH of this solution should be approximately 4.1 since

$$\text{pH} = \frac{(\text{p}K_a + \text{pHA})}{2}$$

So, to arrive at the appropriate pH, some of the $H_2PO_4^-$ must be converted to HPO_4^{2-} by the addition of base. How much depends on the ratio described by the Henderson-Hasselbach equation:

$$pH = pK_a + \log \frac{[A^-]}{[HA]}$$

$$7.3 = 7.2 + \log \frac{[HPO_4^{2-}]}{[H_2PO_4^-]}$$

$$0.1 = \log \frac{[HPO_4^{2-}]}{[H_2PO_4^-]}$$

$$10^{0.1} = \frac{[HPO_4^{2-}]}{[H_2PO_4^-]}$$

$$1.26 = \frac{[HPO_4^{2-}]}{[H_2PO_4^-]}$$

To get the appropriate pH there must be 1.26 times as much HPO_4^{2-} in solution as $H_2PO_4^-$.

$$HPO_4^{2-} + H_2PO_4^- = 0.1 \text{ M}$$

$$1.26 \text{ } H_2PO_4^- = HPO_4^{2-}$$

$$H_2PO_4^- + 1.26 \text{ } H_2PO_4^- = 0.1 \text{ M}$$

$$2.26 \text{ } H_2PO_4^- = 0.1 \text{ M}$$

$$H_2PO_4^- = 0.044 \text{ M}$$

$$HPO_4^{2-} = 0.056 \text{ M}$$

(1 L) $\times$ (0.056 mol L^{-1} HPO_4^{2-}) = 0.056 mol $H_2PO_4^-$ that must be converted to HPO_4^{2-} by titration with NaOH

$$0.056 \text{ mol NaOH}/1 \text{ M NaOH} = 0.056 \text{ L NaOH}$$

56 mL of 1 M NaOH are required for the titration.

To make the buffer, dissolve 13.6 g of KH_2PO_4 in a small amount of water, add 60 mL of 1 M NaOH, and dilute to 1 L with additional water.

Reference

Segal, I. H. (1976). *Biochemical calculations* (2nd ed., p. 441). New York: John Wiley & Sons.

Sample Questions

pH, Acids, and Bases

1. For the following mixtures, identify the conjugate acid and base.
   A. $H_2O + HCl$
   B. $H_2O + H_2PO_4^-$
   C. $NH_3 + H_2O$
   D. $CO_3^{2-} + H_2O$
2. What is the normality (N) of the following solutions?
   A. 0.5 M HCl
   B. 3×10^{-5} M H_2SO_4
   C. 10^{-3} M HNO_3
   D. 4 M NH_3
   E. 1×10^{-6} M H_2CO_3
   F. 1×10^{-4} M H_3PO_4
   G. 1×10^{-2} M Ca^{2+}
   H. 4×10^{-3} M Al^{3+}
   I. 1×10^{-7} M Na^+
3. What is the pH of the following solutions?
   A. 5 M HCl
   B. 2×10^{-5} M H_2SO_4
   C. 4 M HNO_3
   D. 1×10^{-5} M HCl
   E. 2×10^{-5} M NaOH
4. What is the H^+ and OH^- concentration in solutions with the following pH values?
   A. 3.7
   B. 6.4
   C. 7.8
   D. 8.7
   E. 11.4
5. What is the final pH of a mixture of 100 mL of 0.3 M NaOH and 200 mL of 0.2 M H_2SO_4?
6. If you add 0.1 g of NaOH to 1.0 L of distilled water at pH 7, what will the final pH be?
7. If concentrated HCl has a density of 1.19 g cm^{-3} (mL) and contains 37.5 percent HCl by weight, how many moles of HCl are in 25 mL of concentrated HCl and what would be the pH if 25 mL were diluted to 300 mL with water at pH 7.0?
8. What is the pH and pOH of a 0.003 M solution of HNO_3, a strong acid?
9. What is the concentration of HCl in an HCl-amended solution that has a pH of 3.4?
10. What is the $[H^+]$ in a 3×10^{-4} N solution of H_2SO_4?

Neutralization

1. How much 0.1 N HCl would be required to titrate 50 mL of 0.25 N NaOH to neutrality? What size of beaker would you need for your titration?
2. How much 0.2 M H_2SO_4 is required to neutralize 50 percent of the OH^- present in 600 mL of 0.2 M NaOH?
3. What is the pH of a solution in which 0.4 M H_2SO_4 is used to completely neutralize 600 mL of 0.3 M KOH?

Weak Acids

1. What is the pK_a of a 0.3 M weak acid solution (HA) that is 5 percent dissociated?
2. How many milliliters of 0.2 M KOH are required to titrate 300 mL of 0.4 M acetic acid to neutrality?
3. What is the final pH of a solution obtained by mixing 150 mL of 0.3 M NaOH with 400 mL of 0.2 M acetic acid, assuming the pK_a of acetic acid is 4.77?
4. If the pH of a borate buffer is 9.2, how much $H_2BO_3^-$ and HBO_3^{2-} are in a 0.3 M buffer that has a pH of 9.5?

Buffers and Buffering

1. How do you prepare 5 L of 0.2 M glycine buffer at pH 10 if your starting ingredients are glycine (75 g mol^{-1}) and 2 M KOH?
2. What is the relative proportion of mono- and dibasic citrate in a 0.3 M citrate buffer at pH 3.76, 4.76, and 5.76?
3. Prepare 2 L of 0.03 M phosphate buffer, pH 3.0, from stock solutions of 0.5 M H_3PO_4 and KH_2PO_4.
4. Prepare 50 mL of 0.05 M phosphate buffer, pH 11, starting with K_2HPO_4 and a solution of 0.3 M KOH.

11

Oxidation, Reduction, and Energetics

OBJECTIVE

After completing this chapter you should be able to

- determine the oxidation state of elements in soil.
- calculate fermentation balances.
- determine spontaneity of chemical reactions.
- balance chemical equations.

Overview

The soil is in a constant state of flux depending on whether it is aerobic or anaerobic. Aeration largely determines whether soil compounds will be in oxidized or reduced states, which can have significant effects on their chemical behavior. Biological activity is also driven by oxidation as organisms move electrons through a variety of metabolic pathways. Being able to determine whether compounds are oxidized or reduced, and whether chemical reactions are chemically or biologically possible is an important skill for the soil scientist.

Oxidation State or Number

There are a few simple rules for determining the oxidation state of elements in soil (Brock & Madigan, 1991).

1. The oxidation state in the elementary form is zero.
2. The oxidation state of an ion is equal to its charge.

3. The sum of oxidation numbers in a neutral molecule is zero.
4. The sum of oxidation numbers in an ion is equal to the charge of that ion.
5. The oxidation state of O in compounds is virtually always -2, and the oxidation state of H is virtually always $+1$.
6. The oxidation state of C in simple compounds is such that when the oxidation state of O and H in those compounds is summed, the oxidation state of C will make the molecule neutral or reflect the charge on that molecule.
7. In oxidation and reduction reactions, there is a balance between oxidized and reduced products.

Example 11–1

What are the oxidation states of H_2, O_2, N_2, and S_8?

Solution

Rule 1. In each case the oxidation state of these elements in their elementary forms is zero.

Example 11–2

What are the oxidation states of Na^+, K^+, Mg^{2+}, S^{2-}, Cl^-, and Al^{3+}?

Solution

Rule 2. $+1$, $+1$, $+2$, -2, -1, and $+3$, respectively. The oxidation state of the ion of an element is equal to its charge.

Example 11–3

What are the oxidation states of water and glucose?

Solution

Rule 3. The oxidation state of neutral molecules is always zero, although the oxidation state of individual atoms in the compound might not be.

Example 11–4

What are the oxidation states of N in NO_3^- and NH_4^+, and S in SO_4^{2-}?

Solution

Rules 4 and 5. $+5$, -3, and $+6$, respectively. The sum of oxidation states is equal to the charge of the ion. The oxidation states of O and H are always given as -2 and $+1$, respectively.

Example 11–5

What is the oxidation state of C in CO_2 and CH_4?

Solution

Rules 5 and 6. +4 and −4, respectively. The oxidation state of C is such that the charge or neutrality of the compound is maintained.

Example 11–6

What is the oxidation state of C in glucose ($C_6H_{12}O_6$)?

Solution

Rules 5 and 6. The overall oxidation state is zero. The sum of oxidation states of H is +12, and the sum of oxidation states of O is −12. Therefore, to remain in a neutral state, the sum of the oxidation states of C in glucose is zero.

Fermentation Balances

During fermentation, a compound may be used as both an electron donor and acceptor; part of the compound being oxidized and part reduced. It is often useful to distinguish between the two. Because there is a balance between oxidized and reduced products in any oxidation/reduction reaction (see Rule 7 above) the loss of electrons by one compound must be balanced by a gain of electrons in other compounds. With respect to C in fermentation, the extent of oxidation in some carbons of a molecule is balanced by the reduction that occurs in other carbons.

Example 11–7

Example 11–6 demonstrated that the overall oxidation state of C in glucose was zero. Show that during the fermentation of glucose to alcohol, there is a balance between oxidized and reduced products.

Solution

The reaction in question is

$$C_6H_{12}O_6 \quad \rightarrow \quad 2CH_3CH_2OH \quad + \quad 2CO_2$$

$$\text{(Glucose)} \qquad \text{(Ethanol)} \qquad \text{(Carbon Dioxide)}$$

The overall oxidation state of C in ethanol is −4 (six H at +1 = +6; one O at −2 = −2; sum = +4, so to remain neutral overall oxidation state of C must be −4). The oxidation state of C in CO_2 is +4 (two O at −2 = −4; to remain neutral the overall oxidation state of the C must be +4). There are two molecules each of ethanol and CO_2, so a total of eight electrons has been transferred, but the loss of electrons that led to the oxidation of C to CO_2 was exactly balanced by the gain of electrons by C to produce ethanol which is now more reduced (−4) than the starting product (0).

Free Energy Changes

It is important to have a sense for how much energy is generated in various oxidation/reduction (redox) reactions, and whether such reactions will occur spontaneously or not (Segal, 1975).

The practical limits of oxidation-reduction potential (E_h) in soil are the reduction of protons at −414 mV

$$2H^+ + 2e^- \rightarrow H_2 \qquad \text{11-1}$$

and the oxidation of water at 816 mV.

$$2H_2O \rightarrow O_2 + 4H^+ + 4e^- \qquad \text{11-2}$$

These are the practical limits because all life on Earth depends on water. So, if the redox potential is too low, all the water turns to H_2, and if the redox level is too high, all the water oxidizes to O_2.

Redox reactions are written as reductions by convention. The critical feature to remember is that it is the magnitude of the standard redox potential E_h (E_0 in standard conditions) that determines which compound will be oxidized and which compound will be reduced in a coupled redox reaction. It is the magnitude of the difference between two redox potentials that determines how much energy can be generated. Some important redox couples and values for E_0 in the soil environment are given in Appendix 3.

For example, the redox potential for the reduction of Fe^{3+} to Fe^{2+} is given as 770 mV, and the redox potential for the reduction of CO_2 to CH_4 is given as −230 mV. If these reactions were coupled, Fe^{3+} would be reduced because it has the higher redox potential (it's a better electron acceptor). The CH_4 would act as the electron donor in this reaction because it has the lower redox potential.

Example 11–8

Write the coupled redox reaction of Fe^{3+} and CH_4 in the environment.

Solution

$$Fe^{3+} \rightarrow Fe^{2+} \qquad E_0 = 770 \text{ mV}$$

$$\underline{CH_4 \rightarrow CO_2 \qquad E_0 = -230 \text{ mV}}$$

$$Fe^{3+} + CH_4 \rightarrow Fe^{2+} + CO_2 \qquad \Delta E_0 = 1000 \text{ mV}$$

Note that the direction of the reaction in the case of the $CO_2 \rightarrow CH_4$ reduction was flipped to reflect what actually happens in this redox reaction in nature. Also note that no attempt has been made to chemically balance the equation.

To convert the magnitude of the redox potential change into some measure of energy, use the following equation:

$$\Delta G = -nF \, \Delta E_0 \qquad \text{11-3}$$

where

ΔG = Gibbs free energy change
ΔE_0 = redox potential change
n = number of electrons transferred
F = Faraday's constant (23,063 cal $V^{-1} e^{-1}$ or 96.5 kJ V^{-1} mol^{-1})

If ΔG is negative, the reaction is spontaneous, and if ΔG exceeds 7.7 kcal (or 32 kJ), there is enough energy for a microbe or cell to potentially use this reaction to synthesize one adenosine triphosphate (ATP). In reality, it takes about twice as much energy to generate ATP, 15.4 kcal (or 64 kJ).

Example 11–9

What is the free energy change of the reaction in Example 11–8?

Solution

The oxidation of CH_4 to CO_2 liberates eight electrons (i.e., the oxidation state of C changes from -4 to $+4$), the $\Delta E_0 = 1000$ mV.

$$\Delta G = -nF \, \Delta E_0$$

$$\Delta G = -(8)(23{,}063 \text{ cal } V^{-1} e^{-1})(1.0 \text{ V})$$

$$\Delta G = 184.5 \text{ kcal}$$

However, there's an inherent problem in the reaction

$$Fe^{3+} + CH_4 \rightarrow Fe^{2+} + CO_2 \qquad \text{11-4}$$

As it is written, it is unbalanced with respect to both charge and elements. This is a physical and chemical impossibility in the soil environment; charges always balance.

To balance the reaction you need to perform three steps:

1. Make sure that the number of electrons transferred is balanced.
2. Make sure the ionic charge is the same on both sides of the equation, whether it is positive, negative, or neutral.
3. Make sure the total number of atoms of each kind is equal on both sides of the equation.

Example 11–10

Write the overall equation for Example 11–8 as a balanced equation.

Solution

$$Fe^{3+} + CH_4 \rightarrow Fe^{2+} + CO_2$$

1. Eight electrons are lost from CH_4, but only one electron is gained by reducing Fe^{3+} to Fe^{2+}, so

$$8Fe^{3+} + CH_4 \rightarrow 8Fe^{2+} + CO_2$$

2. The charge on the left side of the equation is $+24$ but is only $+16$ on the right side of the equation. This is unbalanced, so $8+$ charges must be added to the right side of the equation, usually as H^+.

$$8Fe^{3+} + CH_4 \rightarrow 8Fe^{2+} + CO_2 + 8H^+$$

3. There are two O's on the right side of the equation, and none on the left. There are also eight H's on the right side of the equation, and only four on the left. Add H_2O to the left side of the equation to complete balancing the reaction.

$$8Fe^{3+} + CH_4 + 2H_2O \rightarrow 8Fe^{2+} + CO_2 + 8H^+$$

References

Brock, T. D., & Madigan, M. T. (1991). *Biology of microorganisms* (6th ed.). Upper Saddle River, NJ: Prentice-Hall.

Segal, I. H. (1975). *Biochemical calculations.* New York: John Wiley & Sons.

Sample Questions

Oxidation State or Number

1. What are the oxidation states of ferric iron and ferrous iron?
2. What is the oxidation state of Mn in MnO_2 and in $MnSO_4$?
3. What is the oxidation state of C in C_2H_6?
4. How does the oxidation state of P change between $H_2PO_4^-$ and HPO_4^{2-}?
5. How many electrons does S gain when SO_4^{2-} is reduced to H_2S?

Fermentation Balances

1. Demonstrate conservation of electrons during fermentation of acetic acid (CH_3COOH) to methane (CH_4) and carbon dioxide (CO_2) during methanogenesis.

2. When glucose ($C_6H_{12}O_6$) is fermented to lactic acid ($C_3H_6O_3$), does the overall oxidation state of the C change?
3. Does glucose ($C_6H_{12}O_6$) fermentation to butyric acid (C_3H_7COOH) form a product that is more reduced or more oxidized?

Free Energy Changes

1. In a coupled redox reaction involving NO_3^- and acetate, which compound will end up being oxidized and which compound will end up being reduced?
2. Can CO_2 reduction to formate be coupled to succinate oxidation to fumarate?
3. What is the ΔE_0 when O_2 reduction is coupled to H_2S oxidation?
4. When glucose oxidation is coupled to O_2 reduction, approximately how many kilocalories of energy can be obtained, and is this a spontaneous reaction?
5. Write a balanced equation for the oxidation of H_2S to SO_4^{2-} by O_2.
6. Write a balanced equation for the oxidation of H_2S to SO_4^{2-} by Fe^{3+}.
7. Which of the following redox coupled reactions are thermodynamically feasible as written? Calculate the ΔG of each feasible reaction assuming a temperature of 298K, pH 7, and 1 atm pressure (i.e., standard conditions).

		E_0
A.	$H_2O \rightarrow \frac{1}{2} O_2 + 2H^+ + 2e^-$	816 mV
	$2Fe^{3+} + 2e^- \rightarrow 2Fe^{2+}$	771 mV
B.	Pyruvate + CO_2 + $2H^+$ + $2e^- \rightarrow$ malate	-360 mV
	$H_2 \rightarrow 2H^+ + 2e^-$	-414 mV
C.	Pyruvate $\rightarrow$ acetate + CO_2 + $2H^+$ + $2e^-$	-700 mV
	$2H^+ + 2e^- \rightarrow H_2$	-414 mV
D.	$2Cu^{2+} + 2e^- \rightarrow 2Cu^+$	150 mV
	$H_2S \rightarrow S_0 + 2H^+ + 2e^-$	-230 mV

12

Kinetics

OBJECTIVE

After completing this chapter you should be able to

- draw a graph of a Michaelis-Menten reaction and identify the critical kinetic parameters.
- linearize the Michaelis-Menten equation.
- write equations for first-order and zero-order processes.
- determine the half-life and mean residence time in first-order and zero-order processes.

Overview

Kinetics deals with factors that determine, measure, and predict the rate of chemical and biochemical processes in soils. Kinetics lets us determine how quickly a process occurs. Knowledge of kinetics and kinetic parameters is essential for a soil scientist to determine such things as the rate of mineralization of materials in soil, and consequently the rate at which nutrients can be delivered to plants. Kinetics is also employed to determine how quickly environmental contaminants will be removed from an environment when a radioactive compound has decayed to the point where it can be safely handled, or how old soil organic matter is. Kinetics will be spread throughout some of the subsequent chapters of this book. For now, we will address the basic kinetic equations that will be employed.

Michaelis-Menten Kinetics

A good place to start looking at kinetics is the classic enzyme kinetics described by the Michaelis-Menten equation.

$$v = \frac{V_{max}[S]}{K_m + [S]}$$

12-1

where

v = velocity of the reaction
V_{max} = maximum velocity of the reaction when the enzyme is saturated with substrate
$[S]$ = substrate concentration
K_m = Michaelis-Menten constant, which is the substrate concentration at $V_{max}/2$

Example 12–1

For the graph below, determine the V_{max}, K_m, and reaction velocity when the substrate concentration is 3 mM.

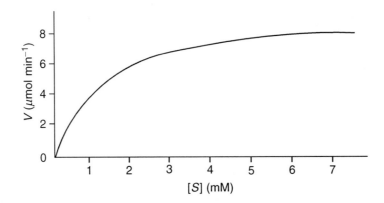

Solution

V_{max} appears to be 8 μmol min^{-1} because the rate does not increase beyond this level. K_m is the substrate concentration at $V_{max}/2$, or 4 μmol min^{-1}, and is approximately 1 mM based on this plot. The velocity of the reaction at 3 mM could be determined directly from the graph but is more accurately determined by the equation.

$$v = \frac{V_{max}[S]}{K_m + [S]}$$

or

$$v = \frac{(8\ \mu\text{mol min}^{-1})(3\ \text{mM})}{1\ \text{mM} + 3\ \text{mM}} = 6\ \mu\text{mol min}^{-1}$$

FIGURE 12–1 The Lineweaver-Burke plot—a linearized version of the Michaelis-Menten equation.

A linear version of the Michaelis-Menten equation is the Lineweaver–Burke plot (Figure 12–1).

$$\frac{1}{v} = \left(\frac{K_m}{V_{max}}\right)\left(\frac{1}{[S]}\right) + \frac{1}{V_{max}}$$ **12-2**

Example 12–2

For the kinetic parameters described in Example 12–1, determine the velocity of the reaction if the substrate concentration was 5 mM by using the Lineweaver-Burke equation.

Solution

It was determined that $V_{max} = 8\ \mu\text{mol min}^{-1}$ and $K_m = 1$ mM, therefore

$$\frac{1}{v} = \left(\frac{K_m}{V_{max}}\right)\left(\frac{1}{[S]}\right) + \frac{1}{V_{max}}$$

$$\frac{1}{v} = \left(\frac{1\ \text{mM}}{8\ \mu\text{mol min}^{-1}}\right)\left(\frac{1}{5\ \text{mM}}\right) + \frac{1}{8\ \mu\text{mol min}^{-1}}$$

$$\frac{1}{v} = 0.15\ \mu\text{mol min}^{-1}$$

$$v = 6.67\ \mu\text{mol min}^{-1}$$

Notice that there are two distinct parts to the curve generated by Michaelis-Menten kinetics. At low substrate concentrations, below the K_m ($K_m >> [S]$), the reaction rate increases as the substrate concentration increases. If you discount the [S] term in the denominator, the Michaelis-Menten equation simplifies to

$$v = \frac{V_{max}}{K_m}[S]$$ **12-3**

In other words, the reaction rate is proportional to a constant (V_{max}/K_m) times the substrate concentration, or as a soil scientist is likely to say, the reaction rate is **substrate dependent.** Adding substrate increases the reaction rate.

Likewise, when there is a high substrate concentration ($K_m <<< [S]$), the terms for [S] cancel, and the Michaelis-Menten equation simplifies to $v = V_{max}$.

In other words, the reaction velocity is a constant (V_{max}), and the reaction is said to be **substrate independent.** Adding more substrate has no effect on the reaction rate.

When describing reaction rates in soil, these two regions define the type of process or equations that can be used to evaluate the reactions. When the reaction rate depends on the substrate concentration, we call this a **first-order process.** When the reaction rate is constant, we call this a **zero-order process.** The mathematics of each is described in the next sections.

First-Order Processes

First-Order Rates

In a first-order process (Segal, 1976)

$$v = -\frac{\Delta[S]}{\Delta t} = k[S]$$

12-4

where

v = velocity or reaction rate
$\Delta[S]/\Delta t$ = substrate consumed per increment of time
k = a constant fraction
$[S]$ = the substrate concentration present at that time

The difference in substrate concentration between when the reaction started (S_0) and when the substrate concentration was measured again (S_t) can be written in the following exponential form:

$$S_t = S_0 e^{-kt}$$

12-5

A plot of this reaction will look like Figure 12–2.

First-order processes proceed slower as time progresses and the substrate runs out. The exponential equation can be written in linear form by taking the natural logarithm (ln) of both sides of the equation.

$$\ln S_t = \ln S_0 - kt$$

12-6

and a plot of the reaction will look like Figure 12–3.

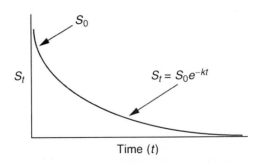

FIGURE 12–2 Plot of a first-order reaction in which the substrate exponentially disappears.

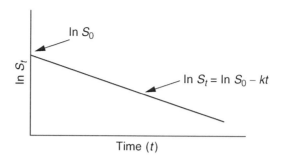

FIGURE 12-3 ln-transformed plot of a first-order reaction in which the substrate concentration exponentially declines.

Example 12–3

What percent of the substrate will remain in a first-order process after 7 days if the first-order rate constant is 0.12 day^{-1}?

Solution

Initially $S_0 = 100\%$, $k = 0.12$ day^{-1}, and the time interval (t) is given as 7 days. Solving by the exponential form gives

$$S_t = S_0 e^{-kt}$$
$$S_t = (100\%)\, e^{-(7 \text{ days})(0.12/\text{day})}$$
$$S_t = 43\%$$

Solving by the linear form gives

$$\ln S_t = \ln S_0 - kt$$
$$\ln S_t = \ln 100 - (7 \text{ days})(0.12 \text{ day}^{-1})$$
$$\ln S_t = 4.605 - 0.84$$
$$\ln S_t = 3.765 = 43\%$$

Half-life Determinations

The time it takes for half the substrate to disappear ($S_0/2$) is called the **half-life** ($t_{1/2}$) and is a very important kinetic property. For a first-order process, the half-life can be calculated by the following equation:

$$t_{1/2} = \frac{\ln 2}{k} = \frac{0.693}{k} \qquad \textbf{12-7}$$

Example 12–4

How long will it take for 50 percent of a substrate to disappear in a first-order reaction if it has a reaction rate constant of 0.4 week^{-1}?

Solution

$$t_{1/2} = \frac{\ln 2}{k} = \frac{0.693}{0.4 \text{ week}^{-1}} = 1.7 \text{ weeks}$$

Mean Residence Time

The **mean residence time** (t_{mrt}) is the mean time it takes for 100 percent of the substrate to disappear. For a first-order reaction, the mean residence time is

$$t_{mrt} = \frac{1}{k} \qquad \text{12-8}$$

Example 12–5

What is the mean residence time of the reaction described in Example 12–4?

Solution

$$t_{mrt} = \frac{1}{k} = \frac{1}{0.4 \text{ week}^{-1}} = 2.5 \text{ weeks}$$

Zero-Order Processes

Zero-Order Rates

In a zero-order reaction, the reaction rate is unaffected by the substrate concentration, or

$$v = -\frac{\Delta[S]}{\Delta t} = k \qquad \text{12-9}$$

In this equation, as before, we write $-\Delta[S]/\Delta t$ (a negative rate) because the substrate is disappearing. The linear form of this equation is $S_t = S_0 - kt$ (Figure 12–4).

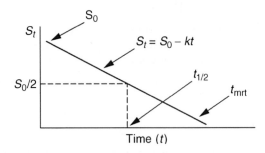

FIGURE 12–4 Plot of a zero-order reaction.

Example 12–6

If a hydrocarbon is decomposing by a zero-order process, how long will it take for 75 percent of the material to disappear if it has a reaction rate constant of 0.1 week^{-1}?

Solution

$$S_t = S_0 - kt$$

$$25\% = 100\% - (0.1 \text{ week}^{-1})(t)$$

$$t = \frac{(100\% - 25\%)}{0.1 \text{ week}^{-1}}$$

$$t = 750 \text{ weeks}$$

Half-life Determinations

The half-life in a zero-order reaction is defined as it was for a first-order reaction, the time it takes for one-half the substrate to disappear. For a zero-order reaction, the equation is

$$t_{1/2} = \frac{S_0}{2k}$$

12-10

Example 12–7

What is the half-life in a zero-order reaction if the starting concentration was 150 ppm and the reaction rate constant was 0.25 week^{-1}?

Solution

$$t_{1/2} = \frac{S_0}{2k}$$

$$t_{1/2} = \frac{150 \text{ ppm}}{(2)(0.25 \text{ week}^{-1})}$$

$$t_{1/2} = 300 \text{ weeks}$$

Mean Residence Time

The mean residence time in a zero-order reaction is the average time it takes for 100 percent of the material to disappear and is calculated as

$$t_{\text{mrt}} = \frac{S_0}{k}$$

12-11

Example 12–8

If the reaction rate constant for a zero-order reaction is 0.33 day^{-1}, how long does it take for a starting substrate concentration of 400 ppm to disappear?

Solution

$$t_{mrt} = \frac{S_0}{k}$$

$$t_{mrt} = \frac{400}{0.33 \text{ day}^{-1}}$$

$$t_{mrt} = 1212 \text{ days}$$

Reference

Segal, I. H. (1976). *Biochemical calculations* (2nd ed.). New York: John Wiley & Sons.

Sample Problems

Michaelis-Menten Kinetics

1. If the V_{max} of a reaction is 15 μmol min^{-1}, what is the velocity of the reaction at its K_m?
2. Given that Michaelis-Menten kinetics apply, what is the velocity of a reaction if [S] = 0.5 mM, V_{max} = 10 μmol min^{-1}, and the K_m = 0.003 mM?
3. If the velocity of a reaction is 5 μmol min^{-1} at its K_m and the K_m of the reaction is 4 mM, at what substrate concentration would you expect V_{max} to occur?
4. Use the Lineweaver-Burke equation to calculate the velocity of a reaction given that the y intercept is 0.2 μmol min^{-1}, the x intercept is -0.04 mM, and the substrate concentration is 5 mM.

First-Order Processes

1. If the first-order rate constant is 0.5 week^{-1}, what will be the concentration of 10 g of substrate after 2 weeks if it decomposes by a first-order process?
2. After 3 weeks the substrate concentration was 25 percent. What is the first-order decomposition rate constant for this reaction?
3. If the initial substrate concentration in a first-order reaction was 200 ppm and the reaction rate constant was 0.33 h^{-1}, what will the substrate concentration be after 15 h?

4. If 1500 ppm of a compound decomposes with a first-order reaction rate constant of 0.25 year^{-1}, how long will it take before the concentration is 750 ppm?

5. What is the mean residence time of oxygen in soil if it is consumed with a k of 0.5 h^{-1}?

Zero-Order Processes

1. For hydrocarbons that decompose via a zero-order process, what will be the concentration if 3000 mg kg^{-1} decompose for 25 weeks with a reaction rate of 0.1 week^{-1}?

2. What is the reaction rate constant for a substrate that was 3500 ppm on day zero and 1200 ppm on day 16?

3. What is the half-life of the material in the question above?

4. There is a high concentration of NO_3^- in a soil sample at the start of a flooding period (10 ppm) and the NO_3^- disappears as described in the following table:

Time (h)	NO_3^- concentration (ppm)
0	10
3	8
6	6
9	4
11	?

What type of reaction is occurring? What is the reaction rate constant? Will any NO_3^- be left to measure after an additional 2 h? If so, how much?

5. A soil sample originally has 120 ppm of an organic substrate. After 8 h it has 100 ppm and after 16 h it has 85 ppm. What type of reaction rate is likely to be represented? What is its reaction constant? How much substrate will be left after an additional 8 h?

6. What is the mean residence time of 1500 ppm benzene if it has a zero-order decomposition rate constant of 0.01 week^{-1}?

13

Stable and Radioactive Isotopes

OBJECTIVE

After completing this chapter you should be able to

- determine the amount of a radioactive isotope remaining after specified intervals.
- calculate delta (δ) values for the most common stable isotopes: ^{13}C, ^{15}N, and ^{34}S.

Overview

Isotopes are elements that have the same number of protons but different number of neutrons in the nucleus. Most elements consist of mixtures of isotopes. Isotopes have the same chemical properties, but because their mass is different and they can be radioactive, isotopes can be distinguished from each other with great sensitivity. Radioactive isotopes are unstable and disintegrate, or decay, into more stable forms by emitting particles such as α (a He nucleus consisting of two neutrons and two protons), β (a high-energy electron), or γ (photons) (Efiok, 1993). These particles differ in the amount of ionizing radiation they have. Over time, radioactive isotopes decrease in abundance. The most common radioactive isotopes a soil scientist may encounter are ^{14}C, ^{3}H, ^{32}P, and ^{35}S. Stable isotopes are not radioactive and constitute a known fraction of a particular element. There are many stable isotopes, but the ones most commonly used in soil science are ^{2}H, ^{13}C, ^{15}N, ^{17}O, and ^{18}O.

Radioactive Isotopes

Radioactive isotopes decay by a simple first-order process (Segel, 1975):

$$\frac{-\delta N}{\delta t} = \lambda N$$

13-1

where

$-\delta N/\delta t$ = number of atoms decaying per unit of time
N = total number of atoms present at a given time
λ = decay constant, which is unique for each radioactive isotope

The decay constant for the most common radioactive isotopes is given in Appendix 4.

This formula can be rearranged into two useful forms

$$N_t = N_0 e^{-\lambda t}$$

13-2

$$\ln N_t = \ln N_0 - \lambda t$$

13-3

where N_t is the amount of radioactive isotope at time t, and N_0 is the starting amount of the radioactive isotope. Equation 13-3 is a linear function in which the slope of the line λ is the decay constant. These equations were discussed in more detail in Chapter 12.

The standard measure of the amount of radioactive decay is the **Curie** (Ci), which is defined as the quantity of radioactive material that gives 2.22×10^{12} disintegrations per minute (dpm). The **specific activity** of a radioactive isotope is the amount of radioactivity per unit of labeled material, or

$$\text{Specific activity} = \frac{\text{radioactivity of sample (Ci)}}{\text{amount of sample}}$$

13-4

Specific activity is usually reported as counts per minute (cpm), which is lower than disintegrations per minute because only a percentage of disintegrations are actually measured by analytical devices.

Example 13–1

Assuming a counting efficiency of 75 percent, what would be the amount of radiation and cpm of a sample that had 3.0×10^7 dpm?

Solution

The amount of radiation is 3.0×10^7 dpm$/2.2 \times 10^{12}$ dpm = $13.636 \ \mu$Ci

$$3.0 \times 10^7 \text{ dpm} \times 0.75 = 2.25 \times 10^7 \text{ cpm}$$

Example 13–2

What is the specific activity of 6 μg of material that contains 50 μCi?

Solution

$$50\ \mu\text{Ci}/6\ \mu\text{g} = 8.33\ \mu\text{Ci}\ \mu\text{g}^{-1} = 8.33\ \text{Ci}\ \text{g}^{-1}$$

As a general rule, there are two things you want to know about radioactive isotopes:

1. For a measured amount of radioactive isotope, how much will remain after a given period?
2. How long did it take for a given amount of radioactive isotope to disappear? This is particularly useful for ^{14}C dating studies.

Example 13–3

A sample of ^{32}P has a specific activity of 5.0×10^7 cpm μmol^{-1}. What will be the specific activity if you have to wait 5 days before using it? The value of λ is obtained from Appendix 4.

Solution

The relevant equation is $N_t = N_0 e^{-\lambda t}$

$$N_t = (5.0 \times 10^7\ \text{cpm}\ \mu\text{mol}^{-1})\ e^{-(0.048/\text{day})(5\ \text{day})}$$

$$N_t = 3.93 \times 10^7\ \text{cpm}\ \mu\text{mol}^{-1}$$

Example 13–4

How much time has elapsed if a sample taken for ^{14}C analysis has only 25 percent of its original ^{14}C content?

Solution

The relevant formula is $\ln N_t = \ln N_0 - \lambda t$

$$\ln 25\% = \ln 100\% - (3.3 \times 10^{-7}\ \text{day}^{-1})(t)$$

$$t = \frac{(\ln 100\% - \ln 25\%)}{(3.3 \times 10^{-7}\ \text{day}^{-1})}$$

$$t = 4{,}200{,}892\ \text{days} = 11{,}509\ \text{years}$$

A very useful term is the **half-life** ($t_{1/2}$), which is the time it takes for half the starting radioactive isotopes to disappear. So, the time it takes for

$$\frac{N_t}{N_0} = \frac{1}{2} \qquad \qquad \boxed{\textbf{13-5}}$$

is equal to the half-life. The half-life for various radioactive isotopes is recorded in Appendix 4.

Half-life can be determined by a very simple equation if you know the decay rate constant.

$$t_{1/2} = \frac{\ln 2}{\lambda} = \frac{0.693}{\lambda}$$

13-6

This is not surprising given that it is exactly the same equation we used to discuss the half-life of first-order reactions in Chapter 12. The only difference is that instead of describing our first-order reaction rate with "k," we are using "λ".

If you know $t_{1/2}$, you can figure out what the λ is for any radioactive isotope. Another useful feature of half-life is that if you know the approximate half-life, you can do back-of-the-envelope calculations (very popular with soil scientists) to quickly determine how much material will remain after a given time, or how long it takes to reduce the material to a given level, simply by repeatedly dividing the starting concentration by 2. The number of times you had to divide times the half-life is the total time it took to reach the desired level.

Example 13–5

What is the half-life of a radioactive compound that has a specific decay constant of 0.012 days?

Solution

$$t_{1/2} = \frac{\ln 2}{\lambda}$$

$$t_{1/2} = \frac{0.693}{0.012 \text{ day}^{-1}} = 58 \text{ days}$$

Example 13–6

The half-life of ^{32}P is approximately 14 days. How many days must you wait before the starting concentration is reduced to <5 percent?

Solution

For a back-of-the-envelope calculation, set up the table below. It shows you have to wait between 56 and 70 days to reach the desired reduction.

Days	Half-lives	Concentration (%)
0	1	100
14	2	50
28	3	25
42	4	12.5
56	5	6.3
70	6	3.1

One of the implications of this procedure, which is useful for soil science studies and for bioremediation, is that for materials decomposing by a first-order process, you have to wait >10 half-lives to ensure that 99.9 percent of the material is removed.

Stable Isotopes

Stable isotopes are typically measured in terms of their atom percent, that is, their proportion of the total isotope pool for that element. If the stable isotope is present in amounts greater than normal, it is reported as **atom percent excess** and the sample is said to be "enriched." On the other hand, it is also possible to remove stable isotopes from a sample so that they are proportionally less than natural abundance. Such samples are said to be "depleted."

Example 13–7

If the natural abundance of ^{15}N in the atmosphere is 0.3663 percent, what would you call a sample with 1 percent ^{15}N?

Solution

This sample is enriched with ^{15}N. It has an atom percent enrichment of

$$1.0000\% - 0.3663\% = 0.6337\%$$

One of the interesting features of stable isotopes is that they are used preferentially in biogeochemical systems, depending on their mass (Schlesinger, 1997). That is to say, the lighter isotopes of an element are slightly more likely to be used in chemical and biogeochemical systems. The end result is that stable isotopes can be used to evaluate things such as (1) global climate change, because ^{18}O is slightly less likely to be used than ^{16}O as the temperature warms; (2) vegetation source, because C_4 plants discriminate between ^{12}C and ^{13}C less than C_3 plants; and (3) denitrification, because $^{14}NO_3^-$ is slightly more likely to be denitrified than $^{15}NO_3^-$.

Soil scientists use something called the **delta** (δ) value to quantify the extent of isotopic fractionation that has occurred. The δ value can be expressed in several ways. For example:

$$\delta\,^{13}C = \left(\frac{^{13}C/^{12}C_{sample} - {}^{13}C/^{12}C_{standard}}{^{13}C/^{12}C_{standard}} \right) \times 1000 \qquad \textbf{13-7}$$

or

$$\delta\,^{13}C = \left(\frac{atom\,\%\,^{13}C_{sample} - atom\,\%\,^{13}C_{standard}}{atom\,\%\,^{13}C_{standard}} \right) \times 1000 \qquad \textbf{13-8}$$

The natural abundance of ^{34}S is 4.5 percent based on a reference standard called the **Canyon Diablo Triolite** (Schlesinger, 1997). As we have seen, the natural abundance of ^{15}N is 0.3663 percent and is based on the ^{15}N abundance in the atmosphere. The natural abundance of ^{13}C in the atmosphere is about 1.1 percent.

Example 13–8

What is the $\delta^{13}C$ of a CH_4 sample that has 1.14 percent ^{13}C?

Solution

$$\delta^{13}C = \left(\frac{\text{atom \% } ^{13}C_{sample} - \text{atom \% } ^{13}C_{standard}}{\text{atom \% } ^{13}C_{standard}} \right) \times 1000$$

$$\delta^{13}C = \left(\frac{1.14 \text{ \% } ^{13}C_{sample} - 1.1 \text{ \% } ^{13}C_{standard}}{1.1 \text{ \% } ^{13}C_{standard}} \right) \times 1000$$

$$\delta^{13}C = 36.4\text{‰}$$

(Note: ‰ rather than % is used to distinguish that we are multiplying the ratio by 1000 rather than 100.)

References

Efiok, B. (1993). *Basic calculations for chemical and biological analysis.* Washington, DC: AOAC International.

Segel, I. H. (1975). *Biochemical calculations* (2nd ed.). New York: John Wiley & Sons.

Schlesinger, W. H. (1997). *Biogeochemistry: An analysis of global change* (2nd ed.). San Diego, CA: Academic Press.

Sample Problems

Radioactive Isotopes

1. For an isotope that has a half-life of 6 years, calculate its decay constant λ.
2. What percent of the original compound in Problem 1 will remain after 15 years?
3. How long will it take for 99 percent of the ^{32}P in a sample to decay?
4. The usefulness of ^{14}C as an indicator diminishes significantly after 99.9 percent has decayed. About how old can an organic sample be before ^{14}C is ineffective as a tracer?

5. Complete the table below:

Isotope	$\lambda(day^{-1})$	$t_{1/2}$	99.9 percent removal from soil
^{14}C	3.3×10^{-7}	5700 years	?
^{13}N	?	13 minutes	?
^{32}P	0.048	?	144
^{33}P	?	25.2 days	256
^{35}S	0.008	?	?

Stable Isotopes

1. What is the atom percent enrichment of a sample that contains 5 percent ^{34}S?
2. How much $^{15}N - NO_3^-$ should be added to 10 g of NO_3-N sample to ensure that it has 1 atom percent enrichment?
3. What is the $\delta^{15}N$ value of a sample that has 0.2 percent ^{15}N?
4. What is the $\delta^{13}C$ value of a sample that has 1.5 percent ^{13}C?
5. What fraction of a sample is ^{34}S if it has a $\delta^{34}S$ of $-2‰$?

Section IV

Problem Solving in Soil Biology

14

Microbial Growth, Yield, and Mortality

OBJECTIVE

In this chapter you will be introduced to some quite simple formulas for making predictions and estimates. After completing this chapter you should be able to

- calculate the number of bacteria growing in ideal conditions.
- mathematically determine the maximum growth rate of bacteria in an environment.
- determine the yield coefficients for bacteria.
- predict the survival rates of bacteria in the environment using simple kinetics.

Overview

In soil microbiology it is very important to be able to determine how fast microorganisms can grow in an environment. This allows you to predict, for example, how rapidly a given process may occur. If your intent is to produce or maintain a given population, it is also critical to know how much substrate, or food, you have to provide to the microbes for growth. During decomposition studies, it is critical to know how much of a compound was truly decomposed, and how much was likely converted into new microbial cells. Finally, in the area of soil and water quality, it is important to know how long potential pathogens can survive in an environment in which they are released. This is vitally important in soils to which human and animal wastes are applied. Unless you know how long it takes for pathogens to be eliminated from the soil, you don't know when it is safe to return to that environment.

Ideal Bacterial Growth

Most of what we know about ideal microbial growth is based on the growth properties of bacteria in the laboratory, and that will be the focus of this chapter. If you grow bacteria in a laboratory, they proceed through a variety of stages as illustrated in Figure 14–1.

Bacteria reproduce almost exclusively by a process called binary fission. In this process one bacterium splits in two, the two bacteria split to make four, the four split to make eight, the eight split to make sixteen, and so on. Under ideal conditions, they divide at uniform rates. This occurs in the **logarithmic or exponential phase** of growth. The time that it takes the bacteria to divide is called the **doubling time** (t_d). During exponential growth, if you know how many bacteria you started with (N_0) and you know how many times the bacteria doubled (called the generation time), you can calculate the total bacteria after a given interval by the following equation:

$$N_t = N_0 \times 2^n \qquad \boxed{\textbf{14-1}}$$

Because the number of generations n is just the total time (t) divided by the doubling time, we can rewrite the equation as follows:

$$N_t = N_0 \times 2^{t/t_d} \qquad \boxed{\textbf{14-2}}$$

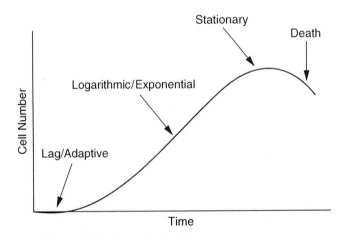

FIGURE 14–1 Stages of microbial growth.

Example 14–1

If you start with eight bacteria and each had a doubling time of 1 h, how many bacteria would be present after 24 h?

Solution

$$N_0 = 8, t = 24, t_d = 1 \text{ so}$$

$$N_t = N_0 \times 2^{t/t_d}$$

$$N_t = 8 \times 2^{24/1}$$

$$N_t = 134{,}217{,}728$$

This is a prodigious number, so microbiologists typically plot growth curves by using logarithms. The logarithmic form of the equation is

$$\ln N_t = \ln N_0 \times \left(\frac{t}{t_d}\right)\ln 2 \qquad \textbf{14-3}$$

Remember that to convert the natural logarithms shown here to values in $\log_{10}$ just divide the natural logarithms by 2.303.

Example 14–2

Convert the equation in Example 14-1 into its linear, logarithmic form, and solve.

Solution

$$N_t = 8 \times 2^{24/1}$$

$$\ln N_t = \ln 8 + \left(\frac{24}{1}\right)\ln 2$$

$$\ln N_t = 2.079 + (24)(0.693) = 18.711$$

$$e^{18.711} = 133{,}685{,}422$$

(The minor differences in final values are because the natural base e is an irrational number.)

If you graph the results, the logarithmic plot of growth during the exponential phase is shown in Figure 14–2. The line intersects the y axis at $\ln N_0$ and the slope of the line is $\ln 2/t_d$.

For any bacteria in a specific environment, the value $\ln 2/t_d$ is a constant called the growth rate constant, and it is given a specific term (μ). The value of

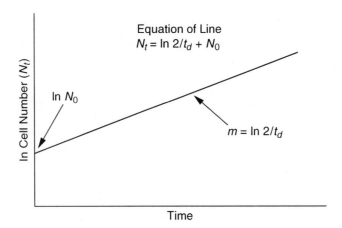

FIGURE 14-2 Exponential growth of ideal bacteria plotted as a linear function.

μ is inversely related to the doubling time, so as the doubling time increases, bacteria grow more slowly as the value of μ declines, and vice versa. One of the powers of knowing μ is that if you know the starting population (N_0), then you can calculate how much the population is changing per unit time in exponential growth by the simple equation $\Delta N = \mu N_0$.

Example 14–3

What is the specific growth rate of a bacterium that has a generation time of 4 h?

Solution

$$\mu = \frac{\ln 2}{t_d}$$

$$\mu = \frac{\ln 2}{4\,\text{h}} \text{(generation time and doubling time } t_d \text{ are synonymous)}$$

$$\mu = 0.173\,\text{h}^{-1}$$

Maximum Growth Rate

The relationship between the specific growth rate of a bacterium and its environment is given by the Monod equation (Figure 14–3).

$$\mu = \frac{\mu_{\text{max}} \times S}{K_s + S}$$

14-4

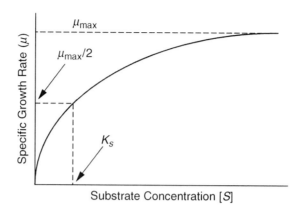

FIGURE 14–3 Growth rate as a function of substrate concentration.

which is exactly analogous to the Michaelis-Menten equation you saw in Chapter 12.

In this case, μ_{max} is the maximum specific growth rate and K_s is the substrate affinity constant—the substrate concentration at which growth is half maximal. When $S >>> K_s$, the equation simplifies to

$$\mu = \mu_{max} \qquad \text{14-5}$$

When $S = K_s$, the equation simplifies to

$$\mu = \frac{\mu_{max}}{2} \qquad \text{14-6}$$

When $S << K_s$, the equation becomes

$$\mu = \left(\frac{\mu_{max}}{K_s}\right) \times S \qquad \text{14-7}$$

which means that the growth rate is proportional to the substrate concentration.

The implications of these equations are that growth rate and metabolic activity of bacteria in the environment will increase if you add growth substrates, particularly if those substrates are limiting in the environment. Conversely, there is a substrate concentration beyond which no further increases in microbial growth and activity can increase; adding more substrate has no effect and is wasteful in terms of materials and money.

Example 14–4

What is the predicted growth rate of a bacterium on glucose if it has a K_s for glucose of 10 μM, a μ_{max} of 1 h^{-1}, and the glucose concentration in the environment is 1 mM?

Solution

$$\mu = \frac{\mu_{max} \times S}{K_s + S}$$

$$\mu = \frac{(1\ h^{-1})(1000\ \mu M)}{(10\ \mu M) + (1000\ \mu M)}$$

$$\mu = 0.99\ h^{-1}$$

Yield Coefficients

The **growth yield coefficient** (Y) is the quantity of biomass formed per unit of substrate consumed. The biomass formed can be in terms of microbial number or in terms of mass (which is more typical). The substrate consumed is usually expressed in terms of mass of substrate consumed, but can also be expressed in terms of other compounds, such as grams of adenosine triphosphate (ATP). Heterotrophic microbes (bacteria that obtain their C for growth from organic C in the environment) typically have yield coefficients ranging from 0.4 to 0.6 g of biomass-carbon per gram substrate C consumed (Silvia et al., 1998). This means that the organisms are 40 to 60 percent efficient in turning their substrate C into new cells. The remaining C is usually lost as CO_2 in aerobic environments and a mixture of CO_2 and CH_4 in anaerobic environments.

Example 14–5

What is the yield coefficient of bacteria that produce 0.25 g of cells for every 0.75 g of C consumed?

Solution

$$Y = \frac{mass\ produced}{C\ consumed}$$

$$Y = \frac{0.25\ g}{0.75\ g} = 0.33\ or\ 33\%$$

Yield coefficients are particularly important terms when it comes to interpreting the results of biomass estimations, which you will examine in Chapter 16, and calculating whether substrates will cause immobilization or mineralization of N in the soil, which you will examine in Chapter 17.

Mortality Rates

The mortality of bacteria added to soil and water is best described as a first-order process, which means it is proportional to the number of cells that are initially added. In its simplest form, which works surprisingly well in most cases, the equation for describing microbial mortality is

$$N_t = N_0 e^{-kt}$$

14-8

or its linear form

$$\ln N_t = \ln N_0 - kt$$

14-9

Knowing the approximate value of k, a first-order rate constant, is extremely valuable because it lets you predict how long pathogenic bacteria may survive in the environment, or at least how long until their populations decline to insignificant numbers (Figure 14–4).

Example 14–6

What concentration of *Shigella* (a bacterial pathogen) will be left in a contaminated water supply after 5 days if the starting population was 1×10^6 cells/100 mL and it has a k for mortality of 0.05 day^{-1}?

Solution

The relevant equation is $N_t = N_0 e^{-kt}$

$$N_t = \left(\frac{1 \times 10^6 \text{ cells}}{100 \text{ mL}}\right) e^{-(0.05/\text{day})(5 \text{ days})} = 778{,}801$$

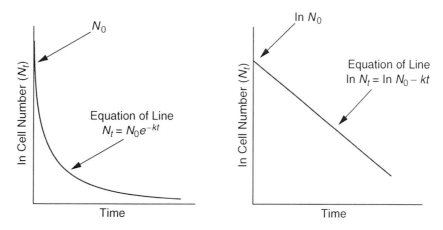

FIGURE 14–4 Mortality of bacteria in soil and water.

Example 14–7

What is the half-life and mean residence time of bacteria that number 1.3×10^5 cells on day 0 and 1.7×10^3 cells on day 6?

Solution

The relevant equation is $N_t = N_0 e^{-kt}$
The linear form will be easier to solve.

$$\ln N_t = \ln N_0 - kt$$

$$\ln 1.7 \times 10^3 = (\ln 1.3 \times 10^5 \text{ cells}) - (k)(6 \text{ days})$$

$$7.438 = 11.775 - (k)(6 \text{ days})$$

$$(k)(6 \text{ days}) = 11.775 - 7.438$$

$$k = (11.775 - 7.4380)/6 \text{ days}$$

$$k = 0.723$$

$$t_{1/2} = \frac{\ln 2}{k} = \frac{\ln 2}{0.723 \text{ day}^{-1}} = 0.96 \text{ days}$$

$$\text{Mean residence time (mrt)} = \frac{1}{k} = \frac{1}{0.723 \text{ day}^{-1}} = 1.38 \text{ days}$$

Remember that from earlier chapters, the half-life ($t_{1/2}$) of any compound or organism that declines by a first-order process is

$$t_{1/2} = \frac{\ln 2}{k} \qquad \textbf{14-10}$$

and its mean residence time is

$$\text{mrt} = \frac{1}{k} \qquad \textbf{14-11}$$

Example 14–8

What is the half-life and mean residence time of E. coli O157:H7 (an important bacterial pathogen in food and water) if it has a k of 0.7 day^{-1}?

Solution

The relevant equations are $\ln 2/k$ and $1/k$ for half-life and mean residence time, respectively.

$$t_{1/2} = \frac{\ln 2}{0.7 \text{ day}^{-1}} = 1 \text{ day}$$

$$\text{mrt} = \frac{1}{0.7 \text{ day}^{-1}} = 1.43 \text{ days}$$

Reference

Silvia, D. M., et al. (1998). *Principles and applications of soil microbiology.* Upper
  Saddle River, NJ: Prentice-Hall.

Sample Problems
Ideal Bacterial Growth

1. If 25 bacteria divided every 35 min for 24 h, what would be the final
   number of cells in the population?
2. If the final population of bacteria is 5×10^5, and there were 10 generations
   of growth, what was the initial population?
3. How many generations of a bacterium will there be after 20 h if it has
   a generation time of 30 min?
4. If the doubling time of a bacterium is 5 h, how many generations will
   there be in 5 days?
5. What is the specific growth rate of the bacterium in Question 4?
6. If the specific growth rate of a bacterium is $5\,h^{-1}$, how many bacteria will
   there be after 20 h if the starting population is 2.5×10^3?
7. Which grows faster, a bacterium with a μ of $0.33\,h^{-1}$ or one with a μ of
   $1\,h^{-1}$?
8. Given a bacterium with a μ of $0.25\,h^{-1}$, how many bacteria will be present
   after 20 h if the initial population was 2?
9. Calculate the starting population if seven generations of a particular bacterium developed to a final population of 1.5×10^7 and the time for
   growth was 10 h. Do you need more information to solve this problem?
10. If you start with two bacteria and the $\mu = 1\,h^{-1}$, how long will it be before
    there are over 1×10^6 cells in this culture?

Maximum Growth Rate

1. If the maximum growth rate of a denitrifier is $0.5\,h^{-1}$, and it has a K_s for
   NO_3^- of $1\,\mu M$, what concentration of KNO_3 should you add to ensure that
   its growth rate will be at μ_{max}?
2. If the measured specific growth rate is $0.01\,h^{-1}$, the μ_{max} is $0.05\,h^{-1}$, and
   the growth substrate concentration at μ_{max} is 25 mM, predict the K_s for this
   organism.
3. What is the maximum specific growth rate of a bacterium that has a K_s of
   $300\,\mu M$ for succinate and is growing at $\mu = 1\,h^{-1}$ when the substrate concentration is 1 mM?

Yield Coefficient

1. What is the yield coefficient of rhizobia that produce 6 mg cells per 20 mg
   sucrose-C consumed?

2. If the yield coefficient of fungi is 60 percent, how much substrate C at a minimum will you need to add to a culture to produce 10 g of cells?
3. How many grams of cells can you expect to produce from 16 g of C if the yield coefficient is 0.1?

Mortality

1. How long must you wait before 99.9 percent of the bacteria are gone if the half-life of the bacteria is 5 days?
2. What is the half-life and mean residence time of bacteria that number 1.3×10^5 cells on day 0 and 1.7×10^3 cells on day 6?
3. How many bacteria will be left after 2 weeks if the starting population was 1.5×10^9 and the k was 0.07 day^{-1}?
4. If the minimum standard for full body contact water is 200 fecal coliforms/100 mL, how many days must you wait for a reservoir to meet this standard if it was contaminated with 2×10^6 bacteria per 100 ml and has a k of 1 day^{-1}?
5. What is the minimum time you must wait for 99.99 percent removal of E. coli from a soil if it has a k of 0.1 $week^{-1}$?

15

Microbial Enumeration

OBJECTIVE

In this chapter you will learn

- basic calculations for estimating microbial populations.

Overview

You can count large organisms in soil by several physical extraction procedures; not so with microorganisms and the smaller soil animals. Even if it is possible to physically extract the organisms, the populations are so high that counting is excessively tedious and time-consuming. For microorganisms, it is also generally impossible to extract all members of the population, and physically impossible to count the enormous numbers that are present in soil. Consequently, methods of counting the smaller organisms in soil (protozoa-sized or smaller) generally rely on some sort of dilution procedure followed by direct counting, plating on selective media, or cultivation to elicit a characteristic population response. These procedures are the topic of this chapter.

Serial Dilution

Serial dilution is one of the most basic procedures in soil biology studies. The principle is very simple (Figure 15–1). A soil sample is suspended in buffer and subsamples are resuspended in additional buffer until an appropriate amount of dilution is achieved. Serial dilution has the same effect as diluting soil in a very large volume of buffer all at once, but is much more convenient and also gives a range of dilutions from which samples can be drawn.

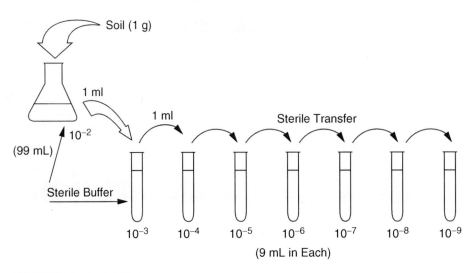

FIGURE 15–1 Serial dilution.

The most critical step in serial dilution is calculating how much soil was originally diluted, and by how much it was diluted. It is important to remember that everything is usually based on grams of oven-dry soil equivalent, even though you typically dilute moist soil. So, you must take into account the volume of water added with soil and the volume contributed by the soil itself. Subsequent dilutions are typically done in 10-fold steps, which simplifies calculations.

The formula for calculating populations after serial dilution is straightforward but requires careful attention to units. First, calculate the **dilution factor** (DF).

$$\frac{\text{g dry soil} \times \text{dilution ratio}_1 \times \text{dilution ratio}_2 \times \ldots \text{dilution ratio}_n}{\text{diluent vol.} + \text{soil vol.} + \text{soil H}_2\text{O}}$$

$$= \frac{\text{g dry soil}}{\text{vol. diluent}} = \text{DF}$$
15-1

The **diluent** is the solution into which the soil is being diluted and serially transferred. The dilution ratio is

$$\frac{\text{volume transferred}}{\text{total volume after transfer}}$$
15-2

Example 15–1

What is the dilution ratio if 5 mL is added to 22 mL?

Solution

The dilution ratio is 5 mL/22 mL + 5 mL = 0.185.

The most complicated step is the initial dilution, because you have to take into account the change in the diluent volume that occurs when you add the soil and the water that is in a moist soil, which also contributes to the total volume.

For example, a prior chapter on soil water (Chapter 7) indicated that 10 g of moist soil at 10 percent gravimetric water content actually contains 0.91 g (or mL) of H_2O and 9.09 g of dry soil ($[0.91/9.09] \times 100 = 10\%$). Furthermore, the 9.09 g of soil has a volume of 3.43 cm^3 (or mL) if we assume a particle density of 2.65 g cm^{-3} (9.09 g/2.635 g cm^{-3} = 3.43 cm^3).

Example 15–2

How much volume does 15 g of moist soil at 35 percent gravimetric water content contribute to 100 mL of diluent?

Solution

$$15 \text{ g moist soil} \div [1 + \text{gravimetric water content (35\%)}]$$
$$= 15 \text{ g} \div 1.35 = 11.1 \text{ g dry soil}$$
$$15 \text{ g moist soil} - 11.1 \text{ g dry soil} = 3.9 \text{ g } H_2O$$
$$11.1 \text{ g dry soil} \div 2.65 \text{ g cm}^{-3} = 4.2 \text{ cm}^3 \text{ soil}$$
$$3.9 \text{ g } H_2O + 4.2 \text{ cm}^3 \text{ soil} = 3.9 \text{ mL } H_2O + 4.2 \text{ mL soil} = 8.1 \text{ mL}$$

The total diluent volume after the addition of soil is 100 mL + 8.1 mL = 108.1 mL. When all is said and done the number of organisms per gram soil is

$$\left(\frac{\text{number of organisms}}{\text{vol. analyzed}} \right) \div DF \qquad \boxed{\textbf{15-3}}$$

Example 15–3

If 10 g of soil at 25 percent moisture content was suspended in 95 mL of water and serially diluted three more times in 10-fold steps, what is the total population if 23 organisms were present when 0.5 mL of the last sample was analyzed?

Solution

The total dilution ratio is 0.001 since each of the three dilution steps was 10-fold, or 1/10. The initial dilution was 10 g/95 mL. However, the 10 g of moist soil contained 2 g of water and only 8 g of soil (2 g H_2O/8 g soil = 25% gravimetric water). The 8 g of soil contributes 3.0 mL to volume (8 g/2.65 g cm^{-3} = 3.0 cm^3 or mL).

So, the total volume of diluent was 95 mL + 2 mL + 3 mL = 100 mL. (This is why so many procedures for serially diluting soil start with an initial dilution of 10 g/95 mL; it's usually close to a 10-fold dilution.) The concentration of soil in each sample analyzed or the dilution factor is therefore

$$\frac{(8 \text{ g} \times 0.001)}{100 \text{ mL}} = 8 \times 10^{-5} \text{ g mL}^{-1}$$

$$\left(\frac{23 \text{ organisms}}{0.5 \text{ mL}}\right) \div 8 \times 10^{-5} \text{ g mL}^{-1} = 5.75 \times 10^{5} \text{ organisms per g soil}$$

Selective Plating

Counting bacteria, fungi, or even viruses typically involves serial dilution of a soil or water sample and then applying the diluted soil or water to a plate of solid media, which is then incubated. For bacteria and fungi, you usually want to have between 20 and 200 colonies growing on the media at the end of incubation. Each colony is assumed to originally represent one bacterial cell, spore, or fragment of fungi. Thus, the organisms you end up counting are termed colony forming units, or CFUs.

Example 15–4

How much will you have to dilute a soil sample that you suspect contains 1×10^6 CFU per gram if you want to plate 1.0 mL and count less than 200 colonies?

Solution

We can estimate the dilution we might need by the following:

$$\frac{200 \text{ CFU mL}^{-1}}{? \text{ g mL}^{-1}} = 1 \times 10^6 \text{ CFU g}^{-1}$$

Solving for "?" gives 2×10^{-4} g mL^{-1} as the final soil concentration. Thus, suspending 10 g of soil in 95 mL of buffer and conducting three more 10-fold dilutions will give an overall dilution of the soil sample of approximately 10,000, which is required for this anticipated soil concentration.

Most Probable Number

Not all soil organisms grow on solid media. In addition, soil biologists sometimes want to enumerate separate physiological groups from among a host of similarly looking organisms. One approach to this problem is a technique called **most probable number** (MPN) enumeration (Alexander, 1982). In MPN enumeration, a soil or water sample is serially diluted and then subsamples from each dilution level are dispensed into replicated tubes of growth media. If at least one viable organism is present, it will grow, or at least cause some characteristic biochemical change in the media that can be detected. The pattern of tubes demonstrating these changes for each dilution is then used to calculate, based on statistical methods, the most probable number of cells that were in the original sample (Figure 15–2).

Once the pattern of positive and negative responses is obtained, a computer program or table is used to calculate the most probable number per gram of soil or milliliter of sample. Most probable number tables are specific to the dilution series and number of replicate tubes that were used for analysis. A table for 10-fold dilutions with five replicate tubes per dilution level is located in Appendix 5. A sample Microsoft Excel program for calculating MPN by spreadsheet is in Appendix 6.

Find the most diluted sample in which all of the inoculated containers give a positive response. Call this dilution p_1. Call the next two highest dilutions p_2 and p_3. From a prepared table of MPN values for the same dilution series and number of replicated samples, read down the column for p_1 until you reach the value for the number of positive samples you observed. Read across for the value of p_2 corresponding to the number of positive samples for p_2. Then read across into the body of the table until you reach the column for p_3 that corresponds to the number of positive samples in that dilution. The number listed is the MPN of organisms in the inoculum you added.

To calculate the MPN in the original sample, divide the tabular MPN by the dilution factor corresponding to p_2. Remember that the dilution factor is

$$\frac{\text{g soil or mL sample}}{\text{vol. of diluent}}$$

and in the MPN table corresponds to the initial dilution times the dilution step.

Example 15–5

In attempting to enumerate nitrifiers from a soil sample, 10 g of field moist soil at 25 percent gravimetric water content was serially diluted in 10-fold steps for each dilution and these dilutions were used to inoculate five replicate tubes of growth media with 1.0 mL from each dilution. The original dilution was 8 g soil/95 mL diluent. The appropriate dilutions for p_1, p_2, and p_3 and the results of the incubation are given in Figure 15–3.

Solution

The number of positive samples for $p_1 = 5$, $p_2 = 4$, and $p_3 = 2$. The corresponding MPN for this sequence in Appendix 5 is 2.2. This number is divided by (initial dilution $\times$ the dilution step) = (8 g/95 mL) $\times$ 10^{-3}. The final value is 2612 nitrifiers per gram soil.

A 95 percent confidence limit can be put on the MPN estimate by multiplying and dividing the calculated MPN by a unique factor for each combination

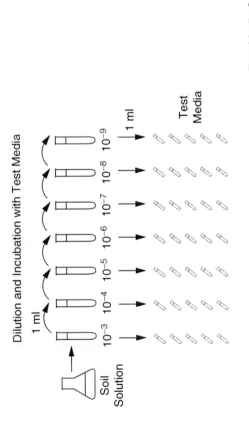

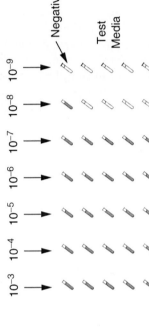

FIGURE 15–2 Schematic diagram of a most probable number procedure.

Initial Dilution Step
8 g Soil/95 mL Buffer

Dilution Step

	10^{-1}	10^{-2}	10^{-3}	10^{-4}	10^{-5}	10^{-6}
Tube 1	+	+	−	+	−	−
Tube 2	+	+	+	−	−	−
Tube 3	+	+	+	+	−	−
Tube 4	+	+	+	−	+	−
Tube 5	+	+	+	−	−	−
	p_1	p_2	p_3			

FIGURE 15–3 Outcome of an MPN enumeration of nitrifiers in a soil sample.

of dilution ratio and number of replicate samples. A table of these factors is located in Appendix 5. As a general rule, unless MPN values differ by a factor of 10, there is not a significant difference between them.

Example 15–6

What is the 95 percent confidence interval for Example 15–5 above?

Solution

It is given that a 10-fold serial dilution with five replications was used. The corresponding factor for this treatment combination is 3.3. Therefore, the 95 percent confidence interval is 2612 nitrifiers per g × 3.3 and 2612 nitrifiers per g ÷ 3.3. The true MPN should therefore lie between 792 and 8620 nitrifiers per g.

Direct Count Calculations

It is well appreciated by most soil microbiologists that an optimistic estimate of the true number of microbes in soil determined by plate counting is only 1 to 10 percent. Furthermore, although the MPN technique overcomes some of the limitations of plate counting, it suffers from an inherent lack of precision. The MPN technique likewise suffers from the limitation that an organism may be present in a sample, yet not be counted because the growth media is wrong.

As long as you can distinguish microbes from a background of organic debris and mineral material, you can use direct microscopy for enumeration. Vital stains, such as acridine orange or fluorescein diacetate, which mark specific cell components such as the cell wall or proteins, can also be used to flag microbes. Direct microscopy will yield a total count of all cells, living and dead, unless the stain distinguishes between the two.

To calculate direct counts, an involved, but relatively simple, equation is used.

$$\frac{\text{cell number}}{W_s \text{ soil}} = \frac{(N_f)(A)}{(a_m)(V_f)\text{DF}}$$

15-4

where

N_f = average number of cells per microscope field
A = filter area (mm² or cm²)
a_m = area of microscope field (mm² or cm²)
V_f = volume filtered (mL)
DF = dilution factor
W_s = dry weight of soil (g)

The area of the microscope field is the actual surface area of the filter that is visible when you look through the lens. Because of magnification, it only represents a fraction of the total surface area of the filter. Consequently, 10 or more microscope fields must be examined to get a reasonable approximation of the average cell density. A special slide called a **stage micrometer** is used to calibrate the field area. A stage micrometer is simply a microscopic ruler incised into a glass slide. Many microscopes are also conveniently equipped with an ocular lens that contains a superimposed grid of squares. Once the length of each square is determined with a stage micrometer, it becomes relatively easy to determine the length of any organism in the microscope field.

Example 15–7

After an initial dilution of 10 g oven-dry soil per 95 mL diluent, the sample is diluted 1000-fold more and 5 mL is dispensed onto a filter. The diameter of the filter is 60 mm and the diameter of the microscope field is 10 μm. If an average of five cells were counted per field, what is the original population?

Solution

The dilution factor is (8 g/95 mL) $\times$ 0.001 = 8.42 $\times$ 10^{-5} g mL^{-1}. The area of the filter which, as you remember from basic geometry is the surface area of a circle, is π (60 mm/2)^2 = 2827 mm². Likewise, the surface area of the microscope field is π (10 μm/2)^2, or π (0.01 mm/2)^2 which is equivalent to 7.854 $\times$ 10^{-5} mm². Entering these values into the appropriate equation gives

$$\frac{\text{cell number}}{\text{g soil}} = \frac{(N_f)(A)}{(a_m)(V_f)\text{DF}} = \frac{(5 \text{ cells})(2827 \text{ mm}^2)}{(7.854 \times 10^{-5} \text{ mm}^2)(5 \text{ mL})(8.42 \times 10^{-5} \text{ g mL}^{-1})}$$

$$= 4.27 \times 10^{11} \text{ cells g}^{-1}$$

Determining the length of microorganisms in soil is particularly applicable to estimating fungal numbers. Because fungi form long hyphae, a more precise measure of the fungi in soil is usually the length rather than the number. By assuming the fungi are approximately the same shape as a cylinder, you can estimate the volume they occupy as well as the length. There is no fundamental difference in the calculation of fungal length as opposed to bacterial number by direct microscopy. The only change is that the average length of fungi is recorded rather than the number of hyphal strands.

Reference

Alexander, M. (1982). Most probable number methods for microbial populations. In A. L. Page et al. (Eds.), *Methods of soil analysis, Part 2. Microbiological and biochemical properties* (2nd ed., pp. 815–820). Madison, WI: Soil Science Society of America.

Sample Problems

Serial Dilution

1. Ten grams of soil was suspended in 95 mL of buffer and serially diluted in 10-fold steps a total of four times. When 1.0 mL was analyzed, there were only two organisms counted. If the starting moisture content was 15 percent, how many organisms were there per gram of soil?
2. Five grams of soil at 20 percent moisture were resuspended in 40 mL of buffer. The suspension was diluted in twofold steps five more times. What was the final soil concentration?
3. Ten grams of soil at 30 percent moisture was suspended in 50 mL of buffer. The suspension was serially diluted in fivefold steps an additional six times. Two milliliters from the final suspension were analyzed and found to have six organisms. What was the original concentration of organisms in the soil?
4. Five grams of oven-dry soil was suspended in 95 mL of water and serially diluted four times in 10-fold steps. What was the final concentration of soil in the last dilution?
5. A microbiologist wanted to dilute exactly 10 g of oven-dry soil in 95 mL of buffer and serially dilute the subsequent suspension 10,000-fold more. If he was starting with moist soil at 25 percent gravimetric water content, what is his final soil concentration?

Plate Counts

1. How many bacteria were present if 0.1 mL of a 10^5 dilution of 10 g of soil yielded 32 bacteria?
2. How many bacteria were initially present if two plates, each inoculated with 1.0 mL of a 10^6 dilution of soil yielded 150 and 75 CFU, respectively?

3. How much will a soil sample have to be diluted if it is suspected to contain 10^4 bacteria per gram soil and you want to inoculate each plate with 0.15 mL of sample?

Most Probable Number

1. Ten milliliters of water was initially diluted in 90 mL of buffer and serially diluted 10-fold in six more steps to determine algae numbers. The p_1, p_2, and p_3 values were 4, 2, and 1, respectively, beginning at the 10^{-3} dilution step. What is the MPN of algae assuming five replicate tubes were used at each dilution?
2. Five grams of moist soil at 25 percent gravimetric water content was diluted in 95 mL of buffer. Three more 10-fold serial dilutions were performed and 2 mL of inoculum was dispensed onto five replicate tubes of growth media. After two weeks the results were recorded. The p_1, p_2, and p_3 values at the 10^{-2}, 10^{-3}, and 10^{-4} dilution steps were 5, 1, and 1, respectively. What is the MPN for this sample?
3. Ten grams of soil at 33 percent moisture were diluted in 95 mL of buffer and examined for the population of mycorrhizal fungi by serial dilution in 10-fold steps and inoculation into replicate containers of sterile soil. The results are shown below.

Dilution step

	10^{-1}	10^{-2}	10^{-3}	10^{-4}	10^{-5}	10^{-6}
Sample 1	+	+	−	−	−	−
Sample 2	+	+	+	−	−	−
Sample 3	+	−	+	+	−	−
Sample 4	+	+	+	−	−	−
Sample 5	+	+	−	−	−	−

What is the MPN for mycorrhizae in the original soil sample?
4. What is the 95 percent confidence interval for an MPN analysis if the calculated MPN was 2.5×10^4 cells per milliliter and the MPN design employed fivefold serial dilutions with seven replications?

Direct Count Calculations

1. One gram of soil was diluted into the equivalent of 1 L, then 2 mL were spread onto the surface of a 60 mm filter. The filter was examined by microscopy and 15 cells were counted per microscope field. Each microscope field had a diameter of 100 μm. What was the total number of cells in the original sample?

2. Ten grams of soil at 33 percent gravimetric water content was diluted in 95 mL water, and four additional fivefold serial dilutions were carried out. Ten milliliters of the resulting suspension were filtered on a 75 mm diameter filter, and 15 cells were counted per microscope field. Each microscope field was 50 μm in diameter. What was the total cell count in the original sample?

3. If the concentration of cells in a water is presumed to be 10^6 mL^{-1} and the sample was diluted 1000-fold before filtering, how much material should be filtered to obtain 25 cells per microscope field, assuming that the filterable area is 5 cm^2 and each microscope field is 100 μm^2?

16

Microbial Biomass

OBJECTIVE

After completing this chapter you should be able to

- understand the basic methods of calculating microbial biomass.
- be able to calculate microbial biomass.

Overview

One of the most fundamental procedures in soil microbiology is determining the mass of the biologically active population. This is a difficult task, because it's impossible to individually isolate all the organisms from soil, nor is it possible to selectively evaluate the biomass of individual populations within soil. Soil microbiologists typically rely on procedures that are indiscriminant in their effect on soil populations but representative of all the populations in soil.

The most common procedures rely on fumigating a soil sample to kill most of the organisms, and then relating biomass to the subsequent evolution of CO_2 when the surviving organisms start to metabolize the dead cells. This is the basis of the chloroform fumigation-incubation procedure for determining microbial biomass, and it is the subject of this chapter.

Chloroform Fumigation Incubation

In the chloroform fumigation-incubation procedure, soil samples are split and one portion is fumigated for 24 h with a saturating amount of chloroform vapor, which kills most, but not all, of the viable microbial biomass. The other portion is not fumigated. After 24 h, the chloroform vapors are removed from the soil, and the fumigated and nonfumigated soil samples are incubated in closed containers for 10 days. During this 10-day incubation, metabolism by the surviving cells occurs, which preferentially focuses on the recently killed microbial biomass. The CO_2 that evolves is trapped in a strong base and measured gravimetrically or by titration, or measured directly by a gas chromatograph. The total CO_2 fluxes from the fumigated and nonfumigated samples are compared, and the difference is related to the microbial biomass that was present before fumigation. One can also calculate microbial N, P, and S by extracting the samples and measuring the inorganic N, P, and S that are produced.

Microbial Biomass Carbon

The calculation of biomass C is straightforward, but has several inherent assumptions.

$$\text{Biomass C} = \frac{(F_c - UF_c)}{K_c} \qquad \textbf{16-1}$$

where

F_c = CO_2-C flush from the fumigated sample
UF_c = CO_2-C flush from the unfumigated control
K_c = fraction of biomass C mineralized to CO_2

The value of K_c ranges from 0.41 to 0.45. It reflects how efficiently microbes use newly released C. The value $1 - K_c$ is analogous to the yield coefficient (Y) for microbial populations, which for most soils is about 60 percent and reflects the dominant contribution of fungi to microbial biomass. The K_c can be determined empirically if necessary, but this is not generally done.

Example 16–1

What is the microbial biomass C of a soil that produces 1 mg g^{-1} CO_2-C from the fumigated sample and 0.2 mg g^{-1} CO_2-C from the unfumigated sample?

Solution

$$\text{Biomass C} = \frac{(F_c - UF_c)}{K_c}$$

$$\text{Biomass C} = \frac{[(1.0 \text{ mg g}^{-1} \text{ CO}_2\text{-C}) - (0.2 \text{ mg g}^{-1} \text{ CO}_2\text{-C})]}{0.41}$$

$$\text{Biomass C} = 1.95 \text{ mg g}^{-1}$$

Microbial Biomass Nitrogen

The calculation of biomass N is essentially the same as that for biomass C.

$$\text{Biomass N} = \frac{(F_N - UF_N)}{K_N}$$

16-2

where

F_N = NH$_4^+$-N flush from the fumigated sample
UF_N = NH$_4^+$-N flush from the unfumigated control
K_N = fraction of biomass N mineralized to NH$_4^+$

To take into account the potential for nitrification during incubation you can base the calculations on the sum of all inorganic N released.

The value of K_N is more variable than that of K_C, and ranges from 0.54 to 0.62. To address this issue, a "floating" value of K_N is sometimes used.

$$K_N = \left[(-0.104)\left(\frac{C_f}{N_f}\right) \right] + 0.39$$

16-3

C_f and N_f are the fluxes of C and N from the fumigated samples, respectively. For C_f/N_f values <6.7, a K_N of 0.54 is sometimes recommended (Horwath & Paul, 1994).

Example 16–2

What is the microbial biomass N of a sample that produces 1 mg g^{-1} NH$_4^+$-N and 2 mg g^{-1} NO$_3^-$-N after fumigation and produces 1.5 mg inorganic N g^{-1} without fumigation for the same incubation period?

Solution

In the absence of any further information, assume that the K_N is 0.54. Therefore

$$\text{Biomass N} = \frac{(F_N - UF_N)}{K_N}$$

$$\text{Biomass N} = \frac{[(1.0 + 2.0 \text{ mg N}) - 1.5 \text{ mg N}]}{0.54}$$

$$\text{Biomass N} = 3.7 \text{ mg g}^{-1}$$

Chloroform Fumigation Extraction

As an alternative to incubating fumigated samples, biomass C and N are also determined by extracting the soil samples with K$_2$SO$_4$ immediately after the 24-h fumigation. The K$_2$SO$_4$ removes C and N that are released from cells lysed by the chloroform. The calculations are essentially the same except that the denominator is replaced by K_{EC} and K_{EN}, respectively, which reflects the efficiency with which the C and N are extracted from soil. These values are 0.35 for K_{EC} and 0.68 for K_{EN}.

Example 16–3

What was the microbial biomass N determined by a chloroform fumigation-extraction procedure that yielded 2 mg N g^{-1} from the fumigated sample and 0.75 mg N g^{-1} from the unfumigated sample after digestion?

Solution

$$\text{Biomass N} = \frac{(F_N - UF_N)}{K_{EN}}$$

$$\text{Biomass N} = \frac{(2.0 \text{ mg g}^{-1} - 0.75 \text{ mg g}^{-1})}{0.68}$$

$$\text{Biomass N} = 1.84 \text{ mg Ng}^{-1}$$

Determining microbial biomass P can be problematic because the inorganic P that is released readily precipitates. To account for this, a second unfumigated control is spiked with a known concentration of inorganic P, usually 25 μg g^{-1}, and extracted to determine the extraction efficiency of any P released (Mullen, 1998).

$$\text{Biomass P} = \left[\left(\frac{25}{C - UF_{EP}} \right) \right] \times \frac{(F_{EP} - UF_{EP})}{K_P} \qquad \text{16-4}$$

where

C = inorganic P extracted from the unfumigated control spiked with 25 μg g^{-1}P
UF_{EP} = inorganic P flush from the unfumigated control
F_{EP} = inorganic P flush from the fumigated sample
K_P = fraction of biomass N mineralized to inorganic P (0.4)

Example 16–4

What is the microbial biomass P of a sample in which 50 μg g^{-1}P is released from a fumigated sample, 15 μg g^{-1} P is released from an unfumigated sample, and 25 μg g^{-1}P is extracted from an unfumigated control spiked with 25 μg g^{-1}P?

Solution

$$\text{Biomass P} = \left(\frac{25\ \mu g\ g^{-1}}{C - UF_{EP}}\right) \times \frac{(F_{EP} - UF_{EP})}{K_P}$$

$$\text{Biomass P} = \frac{(25\ \mu g\ g^{-1})}{(25\ \mu g\ g^{-1} - 15\ \mu g\ g^{-1})} \times \frac{(50\ \mu g\ g^{-1} - 15\ \mu g\ g^{-1})}{0.4}$$

$$\text{Biomass P} = 218.75\ \mu g\ g^{-1}$$

References

Horwath, W. R., & Paul, E. A. (1994). Microbial biomass. In R. W. Weaver (Ed.), *Methods of soil analysis, part 2. Microbiological and biochemical methods* (pp. 753–773). Madison, WI: Soil Science Society of America.

Mullen, M. D. (1998). Transformations of other elements. In D. M. Sylvia et al. (Eds.), *Principles and applications of soil microbiology* (pp. 369–386). Upper Saddle River, NJ: Prentice-Hall.

Sample Problems

1. After fumigation and incubation, 5 mg CO_2-C was measured in a fumigated sample and 1 mg CO_2-C was measured in an unfumigated sample. If the starting amount of soil in each was 25 g, what is the microbial biomass C per gram soil?

2. The CO_2-C flux from a fumigated sample was 7 mg g^{-1} while the N flux was 1 mg g^{-1}. The CO_2-C and N fluxes from the unfumigated controls were 0.9 and 0.09 mg g^{-1}, respectively. What is the apparent microbial biomass C and N?

3. A chloroform fumigation-extraction procedure was performed and 10 mg C g^{-1} was recovered along with 5 $\mu g\ g^{-1}$ NH_4^+ and 10 $\mu g\ g^{-1}$ NO_3^-. If the amount of C and inorganic N extracted from the unfumigated controls was 1 mg C g^{-1} and 0.05 $\mu g\ g^{-1}$ inorganic N, respectively, what is the microbial biomass C and N in this soil?

4. What is the microbial biomass P of a sample that yielded 15 $\mu g\ g^{-1}$ inorganic P after extraction of a fumigated sample, 5 $\mu g\ g^{-1}$ P after extraction of an unfumigated sample, and 20 $\mu g\ g^{-1}$ P after extraction of an unfumigated control spiked with 25 $\mu g\ g^{-1}$ P?

17

Mineralization and Immobilization Rates

OBJECTIVE

After completing this chapter, you should be able to

- write equations for mineralization reactions in soil.
- calculate mineralization rates using sequential and simultaneous models.
- determine the potential for N immobilization in soils receiving a variety of organic substrates.

Overview

Being able to predict the rate at which compounds will decompose or mineralize in the environment is a powerful tool for the soil scientist. Fortunately, the decomposition of many substances can be predicted by relatively simple zero- and first-order reaction rates which were introduced in Chapter 12. In this chapter, we will expand on that discussion to include mineralization of complex substrates as well as examine how the composition of complex substrates can affect N immobilization.

Mineralization

Mineralization of organic compounds in soil is generally predicted by using first-order reaction rates

$$S_t = S_0 e^{-kt}$$

17-1

and its linear form

$$\ln S_t = \ln S_0 - kt \qquad \boxed{17\text{-}2}$$

where

S_t = amount of substrate remaining at any given time after incubation
S_0 = starting substrate concentration
k = first-order reaction rate constant
t = time

Example 17–1

How much cellulose will remain in soil if 100 g is incubated for one week, and it has a first-order decomposition rate of 0.08 day^{-1}?

Solution

$$S_t = S_0 e^{-kt}$$

$$S_t = (100 \text{ g}) \, e^{-(0.08/\text{day})(7 \text{ days})}$$

$$S_t = 57 \text{ g}$$

Soil scientists are just as likely to want to know how much material mineralized as they are how much remains, and this requires a small modification in the formula.

$$S_m = S_0 (1 - e^{-kt}) \qquad \boxed{17\text{-}3}$$

$$\ln S_m = \ln S_0 (1 - kt) \qquad \boxed{17\text{-}4}$$

Everything is as defined before except that S_m represents the amount of material mineralized.

Example 17–2

Using the information from Example 17–1, calculate the amount of cellulose mineralized in 7 days.

Solution

$$S_m = S_0 (1 - e^{-kt})$$

$$S_m = (100 \text{ g})(1 - e^{-(0.08/\text{day})(7 \text{ days})})$$

$$S_m = 42.9 \text{ g}$$

Most substrates added to soil are mixtures of compounds. If you know the chemical composition and the reaction rate constant for each component, then the amount mineralized is equal to the sum of the amount of each fraction mineralized.

$$S_m = [S_1/S_0](1 - e^{-k1t}) + [S_2/S_0](1 - e^{-k2t}) + \cdots + [S_n/S_0](1 - e^{-knt}) \quad \textbf{17-5}$$

where

$$S_m = \text{amount of substrate mineralized}$$
$$S_1/S_0 = \text{percent of fraction 1}$$
$$S_2/S_0 = \text{percent of fraction 2}$$
$$S_n/S_0 = \text{percent contribution of the } n\text{th fraction}$$
$$k_1, k_2 \ldots k_n = \text{reaction rate constants for each individual fraction}$$

S_1/S_0, etc., can also be expressed directly as mass of each fraction if this is known.

Example 17–3

If a 250 g sample of wheat straw is composed of 12 percent sugars, 60 percent cellulose and hemicellulose, and 28 percent lignin with first-order decomposition rate constants of 0.2, 0.08, and 0.01 day^{-1}, respectively, how much straw will remain after a 3-week incubation period?

Solution

$$S_t = (\text{sugars})\, e^{-(0.2/\text{day})(21\text{ days})} + (\text{cellulose})\, e^{-(0.08/\text{day})(21\text{ days})} + (\text{lignin})\, e^{-(0.01/\text{day})(21\text{ days})}$$

$$S_t = (30\text{ g})\, e^{-(0.2/\text{day})(21\text{ days})} + (150\text{ g})\, e^{-(0.08/\text{day})(21\text{ days})} + (70\text{ g})\, e^{-(0.01/\text{day})(21\text{ days})}$$

$$S_t = (30\text{ g})(0.015) + (150\text{ g})(0.186) + (70\text{ g})(0.81\text{ g})$$

$$S_t = 85.05\text{ g}$$

Examples of first-order reaction rate constants for a variety of compounds are in Appendix 7.

Example 17–3 demonstrates a **simultaneous model** for describing mineralization. Each fraction of the substrate decomposes simultaneously although not all of them decompose quickly. You'll notice that only 13.3 g of lignin decomposed during this period.

Because some compounds contain fractions that are preferentially used by soil microbes during decomposition, an alternative model, the **sequential model,** is sometimes used to describe decomposition. In the sequential model, the most readily decomposable substrates are used first, and the decomposition of other substrates does not begin until almost all of the readily decomposed fractions have vanished.

As a general rule you can do a pretty good job of describing the mineralization of complex substrates by using a two-component model in which the substrate is partitioned into a **fast pool** and a **slow pool.** The fast pool represents the fractions that are soluble and have low molecular weight, while the slow pool represents fractions that have high molecular weight, low solubility, and chemical composition that make them resistant to decomposition—compounds such as lignin.

Biosolids, plant residues, and animal wastes are nicely described by two-component models. For example, N mineralization from poultry litter can be modeled using a two-component model with mineralization rate constants of 1.2 to 3.4 day^{-1} for the fast pool (consisting mostly of uric acid) and 0.036 day^{-1} for the slow pool, with the starting N concentration in both pools about the same, 300 to 350 g kg^{-1} of organic N.

Example 17–4

Five megagrams poultry litter, which contained 42 g kg^{-1} total N (TN), was applied to a 1 ha field. The concentration of TN that was rapidly mineralizing N was 340 g kg^{-1} organic N, and the concentration of TN that was slowly mineralizing N was 309 g kg^{-1} organic N. The reaction rate constants for the fast (k_f) and slow (k_s) pools were 1.2 and 0.04 day^{-1}, respectively. How much N will mineralize in 60 days from the organic N pool?

Solution

The total N added to the soil was (42 g kg^{-1})(5000 kg) = 210 kg N.

$$[42 \text{ g kg}^{-1} = 4.2\% \text{ of 5 Mg (5,000 kg)} = 0.21 \text{ Mg (210 kg N)}]$$

The N in the rapidly mineralizing pool is (340 g kg^{-1})(210 kg) = 71.4 kg and the N in the slowly mineralizing pool is (309 g kg^{-1})(210 kg) = 64.9 kg. The appropriate equation is

$$N_m = N_{0F}(1 - e^{-k_f t}) + N_{0S}(1 - e^{-k_s t})$$

$$N_m = (71.4 \text{ kg})(1 - e^{-(1.2/day)(60 \text{ days})}) + (64.9 \text{ kg})(1 - e^{-(0.04/day)(60 \text{ days})})$$

$$N_m = 71.4 \text{ kg} + 59.0 \text{ kg} = 130.4 \text{ kg N}$$

Immobilization

One of the most common problems with the addition of organic residue to soil is the potential for N immobilization. If insufficient N is available in soil or the residue, it can slow mineralization. As a general rule, the C/N ratio at which N immobilization occurs is 30/1. It is also useful to know how much N you might have to add to a substrate to prevent immobilization from occurring.

Again, as a general rule, you can assume that the C/N ratio of the microbial biomass is approximately 8, which reflects that the microbial biomass is composed of about 2/3 fungi and 1/3 bacteria. The overall yield coefficient (Y) for the microbial population is about 0.4, which means that 40 percent of the C mineralized will be converted into biomass and the rest lost as CO_2. This assumes an aerobic environment; anaerobically, the yield coefficient will be closer to 0.1.

What does the C/N ratio of a substrate have to be to prevent immobilization under these assumptions (Myrold, 1998)? Let's assume 100 g of substrate C is available.

- If $Y = 0.4$, then 100 g of substrate will be converted into 40 g of biomass C and 60 g of CO_2-C.
- 40 g biomass C ÷ C/N of 8 = 5 g N needed.
- 100 g substrate C ÷ 5 g biomass N needed = C/N of 20 to prevent immobilization.

Example 17–5

Two hundred grams of a substrate C with a C/N of 75 is added to 1 kg of soil. Assume the yield coefficient for the microbial biomass in this soil is 0.5 and that the microbial biomass C/N is 8. If the soil already contains 10 ppm inorganic N, will immobilization occur, and how much N would need to be added to ensure complete mineralization of the substrate?

Solution

The substrate contains 200/75 = 2.7 g N. With a yield coefficient of 0.5, 100 g of substrate will be turned into microbial biomass C, and 100 g will be respired. The microbial biomass requires 100 g/8 = 12.5 g of N to fully develop. The soil only provides 10 mg N (which is not very much). Not only will this N be immobilized, but the lack of N in the starting material will ensure that not all of it mineralizes unless 12.5 − 2.71 = 9.79 g of additional N are added.

Reference

Myrold, D. D. (1998). Transformations of nitrogen. In D. M. Silvia et al. (Eds.), *Principles and applications of soil microbiology* (pp. 259–294). Upper Saddle River, NJ: Prentice-Hall.

Sample Problems

Mineralization

1. What is the half-life of 2-methylnaphthalene if it has a first-order decomposition rate constant of 25.6 year^{-1}?
2. In a zero-order decomposition process $C_t = C_0 - kt$. If the starting concentration of a waste (C_0) was 150 ppm, and it decomposed at a constant rate

of 0.08 day^{-1}, how many days would it take for 50 percent of the material to disappear?

3. For the data tabulated below, which reaction is proceeding by a first-order process and which reaction is proceeding by a zero-order process? What is the half-life of each compound?

Time (h)	ppm A	ppm B
0	282	333
1	256	287
2	230	247
4	179	183
8	77	100
16	0	30
32	0	3

4. How many days will it take for 1000 ppm *p*-xylene to decline by 99 percent if it has a first-order decomposition rate constant of 0.064 day^{-1}?

5. Graphically show how many days it will take for 1000 ppm *m*-xylene to be reduced by 99 percent if it has a half-life of 7.5 days and also decomposes by a first-order process.

6. The decomposition of straw was assumed to take place by a simultaneous two-component model in which 67 percent of the straw was easily mineralizable with a *k* of 0.019 day^{-1} and the remainder was slowly mineralizable with a *k* of 7.1 × 10^{-4} day^{-1}. How much straw will be available after 30 days if 1 kg of this material is added to soil?

Immobilization

1. Crop residue (50% C, 1.5% N) is chopped and incorporated into the soil. If the average microbial biomass C/N for this soil is 12 and the biomass has a yield coefficient of 0.5 during decomposition, will net N mineralization or immobilization occur?

2. For soil with a microbial biomass consisting of 60 percent fungi and 40 percent bacteria with yield coefficients of 0.5 and 0.25, respectively, calculate the necessary C/N ratio of a substrate that will prevent N immobilization if it is incorporated into the soil. It is also given that the C/N of fungal cells is 10 and that the C/N of bacterial cells is 8.

3. How much additional N must be added to decompose 100 g of an organic substrate with a C/N of 100 added to 1 kg of soil if the microbial yield coefficient is 0.6, the microbial C/N is 12, and the soil has 50 ppm inorganic N?

4. In Question 3 above, how much of the substrate can be mineralized if no additional N is forthcoming?
5. During the mineralization of grass roots in soil, the following observations were made: The roots had a C/N of 22 and were composed of 80 percent nonlignin C, with a k of 1.6 year^{-1} and 20 percent lignin with a k of 0.001 year^{-1}. If the only source of N for mineralization came from root mineralization, and there was 100 g of root biomass per square meter, how much net mineralization will occur if you assume the yield coefficient of the microbial community is 0.6 and the microbial biomass C/N is 8? (adapted from Myrold, 1998)

18

Respiration and Gas Fluxes

OBJECTIVE

In this chapter you will learn

- how to calculate CO_2 production rates using a base trap and gas chromatograph.
- how to quantify gas fluxes from field environments.

Overview

Gas production during microbial activity in soil is one of the ways in which soil scientists detect the presence of activity, the type of activity, and the magnitude of activity in soil. One of the best indicators of microbial activity in soil is soil respiration, which is the measurement of CO_2 evolved from soil over a given period; the higher the respiration, the greater the activity, and presumably, the better the soil. One can also look for evidence that anaerobic conditions exist in soil by looking for evidence of CH_4 production. Likewise, there is a direct correlation between the aeration status of a soil and denitrification; the worse the aeration, the more likely denitrification is to occur, which can be evaluated by measuring one of its gaseous intermediates—nitrous oxide (N_2O).

In this chapter we will look at two ways of evaluating gas fluxes. First, the use of base traps to measure soil respiration. Second, the measurement of gas flux from intact cores and static soil covers.

Respiration by Titration

Respiration produces CO_2 in aerobic systems, so the greater the CO_2 produced, the greater the respiration. The easiest and cheapest way of measuring CO_2 flux is to trap the CO_2 evolved from a known mass, volume, or area of soil in a strong base such as NaOH and titrate the resulting solution with a known concentration of acid (Zibilske, 1994). The basic chemistry of the reaction is as follows:

$$CO_2 + 2Na^+ + 2OH^- \leftrightarrow CO_3^{2-} + 2Na^+ + H_2O \qquad \textbf{18-1}$$

For every mole of CO_2 trapped by the base, 2 mol of OH^- are consumed. After the incubation period, excess $BaCl_2$ is added to the base trap to precipitate the CO_3^{2-} and prevent it from reacting during the subsequent acid titration.

$$BaCl_2 + Na_2CO_{3(soluble)} \leftrightarrow BaCO_3 + 2NaCl \qquad \textbf{18-2}$$

Finally, the residual OH^- is titrated with an acid (such as HCl) to determine how much remains.

$$HCl + Na^+ + OH^- \leftrightarrow NaCl_{(soluble)} + H_2O \qquad \textbf{18-3}$$

The difference between the HCl required to titrate a control with no CO_2 added, and that required to titrate a sample trapping CO_2, is proportional to the amount of CO_2 that was produced.

$$mg\ CO_2 = (B - V) \times (N \times E) \qquad \textbf{18-4}$$

where

B = volume of standard acid required to titrate the trap solution of the control

V = volume of standard acid required to titrate the trap solution from the sample flasks to neutrality

N = normality of the acid (usually milliequivalents per milliliter)

E = equivalent weight of CO_2; 22 if reported as mg CO_2 and 6 if reported as mg CO_2-C

Example 18–1

How much CO_2 was produced from a 10 g soil sample if it required 5 mL of 1 N HCl to titrate a base trap containing 10 mL of 2 N NaOH and 19 mL of 1 N HCl to titrate a control with and equal volume of 2 N NaOH?

Solution

$$mg\ CO_2 = (B - V) \times (N \times E)$$

$$mg\ CO_2 = (19 - 5) \times (1 \times 6)$$

$$mg\ CO_2\text{-C} = \frac{84\ mg\ CO_2\text{-C}}{10\ g} = 8.4\ mg\ CO_2\text{-C}\ g^{-1}$$

Gas Fluxes—Soil Cores

Gas fluxes from soil are typically measured in either intact soil cores or by static soil covers. The gas flux from cores is calculated as

$$Q = \frac{(M_2 - M_1)}{t_2 - t_1}$$ **18-5**

where

$$Q = \text{gas flux}$$
$$M_2 \text{ and } M_1 = \text{total gas produced at two sampling intervals}$$
$$t_2 \text{ and } t_1 = \text{two sampling intervals}$$

The gas produced is calculated as

$$M = C_s(V_g + V_l\alpha)$$ **18-6**

where

$$C_s = \text{gas concentration in the gas phase (mass/volume)}$$
$$V_g = \text{total gas volume in the core}$$
$$V_l = \text{volume of liquid in the core}$$
$$\alpha = \text{Bunsen coefficient (the milliliters of gas absorbed per mL } H_2O)$$

Values for the Bunsen coefficient for various gases are given in Appendix 8 (Tiedje, 1994).

Several calculations are required to come up with the values for this equation. The gas concentration is usually measured in terms of parts per million (microliters per liter [$\mu L/L$]). To convert microliters per liter to micrograms per liter ($\mu g/L$)

$$\frac{\mu L}{L} \times \frac{1\ \mu mol}{22.4\ \mu L} \times \frac{?\ \mu g}{\mu mol}$$ **18-7**

Example 18–2

What is the concentration of CO_2 on a mass/volume basis if the measured CO_2 concentration is 500 ppm?

Solution

$$(\mu L\ L^{-1}) \times (1\ \mu mol/22.4\ \mu L) \times (\mu g\ \mu mol^{-1})$$

$$(500\ \mu L/L\ CO_2) \times \left(\frac{1\ \mu mol\ CO_2}{22.4\ \mu L\ CO_2}\right) \times (44\ \mu g\ CO_2/\mu mol\ CO_2)$$

$$982\ \mu g\ CO_2\ L^{-1}$$

The volume of gas (V_g) is the sum of the gas volume in the headspace and the air-filled pore space. The air-filled pore space is the difference between the total core porosity and the volume of liquid. The total porosity is

$$\left[1 - \left(\frac{\text{Bulk density}}{\text{Particle density}}\right)\right] \times \text{core volume} \qquad \textbf{18-8}$$

while the volume of liquid is determined by the difference in core weight before and after oven drying.

Example 18–3

What is the gas volume in a soil core 5 cm in diameter and 10 cm high that has a bulk density of 1.2 g cm^{-3} and weighs 84 g when wet and 74.1 g when dry?

Solution

The total core volume is $(10 \text{ cm}) \pi (5/2)^2 = 196.4 \text{ cm}^3$.
The core porosity is $[1 - (1.2 \text{ g cm}^{-3}/2.65 \text{ g cm}^{-3})] \times 100 = 54.7\%$.
The volume of water is $84.1 \text{ g} - 74 \text{ g} = 10.1 \text{ g or cm}^3$.
$196.4 \text{ cm}^3 \times 54.7\% = 107.4 \text{ cm}^3$ total pore space $- 10.1 \text{ cm}^3$ water $= 97.3 \text{ cm}^3$ air-filled pore space and 10.1 cm^3 water-filled pore space.

Example 18–4

For the soil core described in Example 18–3, the N_2O concentration measured in the headspace (100 cm^3) was 560 ppm. What is the mass of N_2O-N present in this sample? Assume the temperature is 25°C.

Solution

$$M = C_s (V_g + V_l \alpha)$$

$$C_s = (560 \ \mu L \ L^{-1}) \times (1 \ \mu mol/22.4 \ \mu L) \times (28 \ \mu g \ N_2O\text{-}N/\mu mol)$$

$$= 700 \ \mu g \ L^{-1} \ N_2O\text{-}N$$

$$V_g = 100 \text{ cm}^3 \text{ headspace} + 97.3 \text{ cm}^3 \text{ air-filled pore space}$$

$$= 197.3 \text{ cm}^3 = 0.197 \text{ L}$$

$$V_l = 10.1 \text{ cm}^3 = 0.0101 \text{ L}$$

$$M = (700 \ \mu g \ L^{-1} \ N_2O\text{-}N)[0.197 \text{ L} + (0.0101 \text{ L} \times 0.544)]$$

$$M = 142 \ \mu g \ N_2O\text{-}N$$

Gas Fluxes—Soil Covers

Another technique for measuring gas fluxes from soil is to use an intact soil cover. Because the depth of soil that actually supplies gas to the soil cover is not known, flux measurements are usually made on an area basis using this equation (Mosier & Klemedtsson, 1994).

$$\text{Gas flux} = \frac{V(C_1 - C_0)^2}{(A)(t)(2C_1 - C_2 - C_0)} \ln \frac{(C_1 - C_0)}{(C_2 - C_1)} \qquad \textbf{18-9}$$

where

V = volume enclosed by the soil cover
C_0, C_1, C_2 = gas concentrations at various sample intervals
t = time interval
A = surface area enclosed

When $C_1 - C_0 = C_2 - C_1$, the equation simplifies to

$$\text{Gas flux} = \frac{V(C_1 - C_0)}{(A)(t)} \qquad \textbf{18-10}$$

The correction factor

$$\ln \frac{C_1 - C_0}{C_2 - C_1}$$

is used to account for the reduction in the gas concentration gradient as the gas accumulates in the soil cover (Mosier & Klemedtsson, 1994).

Example 18–5

The following data were collected during a soil cover experiment:

Time (min)	ppm CO_2
0	330
30	430
60	520

If the soil cover had a volume of 750 mL and covered 125 cm² of soil, what is the calculated CO_2 gas flux?

Solution

$$\text{Gas flux} = \frac{V(C_1 - C_0)^2}{(A)(t)(2C_1 - C_2 - C_0)} \ln \frac{(C_1 - C_0)}{(C_2 - C_1)}$$

$$\text{Gas flux} = \frac{0.75 \text{ L} (430 - 330)^2}{(0.0125 \text{ m}^2) (30 \text{ min}) (860 - 520 - 330)} \ln \frac{(430 - 330)}{(520 - 430)}$$

$$\text{Gas flux} = 0.2 \text{ ppm CO}_2 \text{ m}^{-2} \text{ min}^{-1}$$

$$= (0.20 \ \mu\text{L L}^{-1}) \times (1 \ \mu\text{mol}/22.4 \ \mu\text{L}) \times (12 \ \mu\text{g CO}_2\text{-C}/\mu\text{mol})$$

$$= 0.107 \ \mu\text{g CO}_2\text{-C m}^{-2} \text{ min}^{-1}$$

References

Mosier, A. R., & Klemedtsson, L. (1994). Measuring denitrification in the field. In R. W. Weaver et al. (ed.) *Methods of soil analysis, part 2. Microbiological and biochemical properties* (pp. 1047–1065). Madison, WI: Soil Science Society of America.

Tiedje, J. M. (1994). Denitrification. In R. W. Weaver et al. (Eds.), *Methods of soil analysis, part 2. Microbiological and biochemical properties* (pp. 245–267). Madison, WI: Soil Science Society of America.

Zibilske, L. M. (1994). Carbon mineralization. In R. W. Weaver et al. (Eds.), *Methods of soil analysis, part 2. Microbiological and biochemical properties* (pp. 835–863). Madison, WI: Soil Science Society of America.

Sample Problems

Respiration by Titration

1. What is the maximum amount of 1 N HCl required to titrate 15 mL of 0.5 N NaOH?

2. Two base traps containing 20 mL of 0.5 N NaOH were each exposed to 5 g of soil for 1 week. After this incubation, one trap was titrated with 0.5 N HCl and the other was titrated with 0.5 N H_2SO_4. The volume of acid required to titrate the first trap was 10 mL and the volume of acid required to titrate the second trap was 4 mL. Assuming that 20 mL and 10 mL were required to titrate the controls for each trap, respectively, what was the CO_2 production in mg CO_2-C g^{-1} for each sample?

3. One hundred grams of soil were incubated with 50 mL of 2 N NaOH. After 1 week, 10 mL of the NaOH trap was titrated with 1 N HCl. A similar amount of NaOH from a control was titrated with the standard acid and required 19 mL. The sample itself required 8 mL of standard acid for titration. What was the mg CO_2-C produced per gram soil?

Gas Fluxes—Soil Cores

A soil core 15 cm tall by 10 cm diameter was removed from a field, sealed in a container, and N_2O was measured in the headspace at 1 and 2 h. The concentration of N_2O was 1.0 ppm at 1 h and 5 ppm at 2 h. The incubation temperature was 25°C and the core was found to have 30 percent gravimetric water content. If the bulk density was 1.1 g cm³, answer the following questions:

1. What was the total volume of gas sampled?
2. What was the concentration of gas in the head space at 1 h?
3. What was the N_2O-N flux from this soil core?
4. What was the N_2O flux per gram soil per day?

Gas Fluxes—Soil Covers

1. The following data were collected during a soil cover experiment that employed a soil cover with a volume of 1 L, covering a surface area of 100 cm². What is the CO_2-C flux m^{-2} h^{-1}?

Time (min)	ppm CO_2
0	430
30	530
60	630

2. The following data were collected during a soil cover experiment that employed a soil cover with a volume of 0.75 L, covering a surface area of 125 cm². What is the N_2O-N flux m^{-2} h^{-1}?

Time (min)	ppm N_2O
0	0.330
60	0.530
120	0.730

3. The following data were collected during a soil cover experiment that employed a soil cover with a volume of 1.0 L, covering a surface area of 75 cm². What is the CO_2-C flux m^{-2} h^{-1}?

Time (min)	ppm CO_2
0	330
60	530
120	630

Section V

Problem Solving in Soil Chemistry, Fertility, and Management

19

pH, Liming, and Lime Requirements

OBJECTIVE

There are few measurements more important or useful in soil science than pH, and there are few practices more likely to increase plant yield than liming or (rarely) acidifying soil to an appropriate pH. In this chapter you will

- review the meaning of pH.
- practice calculations to determine the liming potential of various compounds.
- calculate lime requirements for soil at various pH.

Overview

Soils are unique physical, chemical, and biological environments. But it is important to recognize that soils are also managed environments—managed by crop selection, fertilization practices, and tillage operations. In terms of soil amendments then, it is critical to be able to calculate how much of an amendment to add to bring about the desired result. This is particularly important in the case of managing pH by liming, which is the topic of this chapter.

pH

It bears repeating that pH is the negative logarithm (base 10) of the hydronium ion (H_3O^+) activity, which is very nearly the same as the H_3O^+ concentration in water.

$$pH = -\log_{10}[H_3O^+] \qquad \boxed{\textbf{19-1}}$$

You are more likely see the hydronium ion represented as H^+, which is what we will do for the rest of this chapter.

A soil with a pH of 7 is considered neutral. A soil in which pH <7 is acidic. A soil in which pH >7 is basic or alkaline. The pH declines when the $[H^+]$ increases, and the pH rises when the $[H^+]$ decreases (Table 19–1).

Example 19–1

What is the $[H^+]$ when the pH is 4?

Solution

$$[H^+] = 10^{-4}\,M$$

Example 19–2

If the $[H^+]$ is 10^{-7}, what is the pH?

Solution

$$pH = 7$$

TABLE 19–1   Relationship of pH to H^+ Concentration

pH	$[H^+]$	Comments
3	0.001 M (10^{-3} M)	Strongly acidic
4	0.0001 M (10^{-4} M)	
5	0.00001 M (10^{-5} M)	
6	0.000001 M (10^{-6} M)	Slightly acidic
7	0.0000001 M (10^{-7} M)	Neutral
8	0.00000001 M (10^{-8} M)	Slightly alkaline
9	0.000000001 M (10^{-9} M)	
10	0.0000000001 M (10^{-10} M)	Strongly alkaline

Liming

Liming is synonymous with raising soil pH, although adding limestone ($CaCO_3$) or lime (CaO) is not the only way to accomplish this. Limestone works on the basis of the following reaction:

$$CaCO_3 + 2H^+ \rightarrow Ca^{2+} + H_2O + CO_2 \text{ (g)} \qquad \boxed{\textbf{19-2}}$$

Various materials have differing capacities to neutralize soil acidity (Table 19–2). For example, 1 ton of hydrated lime has the equivalent neutralizing capacity of approximately 1.4 ton of calcitic limestone. The effectiveness of liming materials is determined by its **effective calcium carbonate equivalence** (ECCE). You may also see this referred to as the **relative neutralizing value** (RNV) or **effective neutralizing value** (ENV). This is really a combination of two factors—the effectiveness of a material (how fine it is) × calcium carbonate equivalence (CCE) (how pure it is).

The smaller the size of the liming material, the more reactive it is. Fineness is measured by the ability to pass through meshes of different sizes—the larger the mesh size, the finer the particle (Table 19–3). Anything unable to pass through a 10 mesh sieve is chemically inert for the purpose of liming.

Liming materials, particularly limestone, are usually not completely pure but contain materials that do not contribute to neutralizing capacity. The CCE for a given liming material is adjusted to reflect its purity.

Example 19–3

What is the CCE of dolomitic limestone that is only 85 percent pure?

TABLE 19–2 Calcium Carbonate Equivalence (CCE) of Various Liming Materials

Type of Lime	Formula	Formula Weight (g mol^{-1})	CCE Range (%)
Calcitic limestone	$CaCO_3$	100.1	80–100
Marl	Impure $CaCO_3$ (Mostly shells)		70–90
Dolomitic limestone	$CaMg (CO_3)_2$	184.4	110
Quick lime (Burnt lime)	CaO	56.1	150–180
Hydrated lime (Slaked lime)	$Ca(OH)_2$	74.1	120–140
Wood ash	Various oxides (e.g., MgO)		30–70

TABLE 19–3 Mesh Size and Relative Liming Effectiveness

Mesh Size	Exclusion Range	Relative Liming Effectiveness (%)
>50	Passes through 50 mesh	100
10–50	Passes through 10 mesh, but retained on 50 mesh	50
<10	Retained on 10 mesh	0

Solution

$$0.85 \times 110\% = 93.5\%$$

It would require 1.07 ton of dolomitic limestone with this purity to have the same neutralizing capacity of 1 ton of calcitic limestone.

To calculate the ECCE, use the following equation:

$$\text{ECCE} = (\%\text{CCE}/100) \times 0.5 \ (\% < 10 \text{ mesh} + \% < 50 \text{ mesh}) \qquad \boxed{\textbf{19-3}}$$

Example 19–4

A supply of calcitic limestone is purchased that is 75 percent pure. Ninety five percent of the material passes through 10 mesh, while only 75 percent passes through 50 mesh. What is the ECCE?

Solution

$$\text{ECCE} = \left(\frac{75\%}{100}\right) \times 0.5 \ (95\% + 75\%) = 63.8\%$$

How do you use the ECCE to calculate lime application rates?

Example 19–5

If the liming recommendation called for 3 ton of calcitic limestone, based on the ECCE from Example 19–4, how much should be added?

Solution

$$\text{Actual lime rate} = \frac{\text{desired rate} \times \text{CCE of liming material}}{\text{ECCE of material applied}}$$

$$\text{Actual lime rate} = 3 \text{ ton acre}^{-1} \times \left(\frac{100\%}{63.8\%}\right) = 4.7 \text{ ton acre}^{-1}$$

Lime Requirements

There are three kinds of acidity that contribute to the total acidity of soil—active acidity, exchangeable acidity, and residual or reserve acidity. Active acidity is what you measure when you measure the water pH of a soil. Exchangeable acidity reflects the exchangeable Al^{3+} and H^+ on clay and organic matter surfaces. However, residual acidity represents the greatest contributor of acidity to soil (1000 to 10,000 times as much as the other two), so liming recommendations are based on neutralizing residual soil acidity. There is an equilibrium between H^+ and Al^{3+} in soil solutions, which contributes to active acidity, and that remaining on the soil exchange surfaces as part of residual acidity, so lime requirements can be based on water pH alone. However, most soil testing services measure the pH in equilibrium with buffer (the buffer pH) and derive liming recommendations accordingly.

In most cases, if the water pH is greater than 6.4, liming is not required because there will often not be any demonstrable increase or decrease in yield. When lime is required, the relationship between water and buffer pH can be used to determine the lime needed, which typically assumes a CCE of 67 to 80 percent (based on the quality and size of most agricultural limestones). An example of a table that can be used to calculate lime requirements is shown in Table 19–4.

TABLE 19–4 **Rate of Agricultural Limestone (tons per acre) Needed to Raise Soil pH to 6.4**

Water pH of Sample	Buffer pH of Sample								If Buffer pH is Unknown
	5.5	5.7	5.9	6.1	6.3	6.5	6.7	6.9	
	Ag. Limestone Rate (tons per acre)								
4.5	7.0	6.0	6.0	5.0	4.0	4.0	3.0	3.0	4.0
4.7	7.0	6.0	6.0	5.0	4.0	4.0	3.0	3.0	4.0
4.9	7.0	6.0	6.0	5.0	4.0	4.0	3.0	3.0	4.0
5.1	7.0	6.0	5.0	5.0	4.0	3.0	3.0	2.0	4.0
5.3	7.0	6.0	5.0	4.0	4.0	3.0	3.0	2.0	3.5
5.5	6.0	5.0	5.0	4.0	4.0	3.0	2.0	2.0	3.0
5.7	6.0	5.0	4.0	4.0	3.0	3.0	2.0	2.0	2.5
5.9	—	5.0	4.0	3.0	3.0	2.0	2.0	1.0	2.0
6.1	—	—	3.0	3.0	2.0	2.0	1.0	1.0	1.5
6.3	—	—	—	2.0	1.0	1.0	1.0	1.0	1.0

Note. 2001–2002 Lime and Fertilizer Recommendations, AGR-1. Lexington, KY: University of Kentucky, Cooperative Extension Service.

Example 19–6

How much lime is required to raise the soil pH to 6.4 if the water pH is 5.1 and the buffer pH is 6.3?

Solution

Read across Table 19–4 on the row containing a water pH of 5.1. Read down the table underneath the buffer pH of 6.3. Where row and column intersect, the value is 4.0. Therefore, the recommendation is for 4.0 ton acre^{-1} of agricultural limestone (assuming a CCE of 67 percent) to raise the pH to 6.4.

Some tables do not require you to use water pH to calculate the lime requirement. They will be based on the buffer pH alone.

In addition to assuming a CCE of 67 percent, or some equivalent value, most lime recommendations assume that the material will be uniformly applied and incorporated to a depth of approximately 6 in (15 cm). The recommendation also assumes that approximately 4 years will be required for the limestone to completely react, although materials such as slaked or burnt lime react much more quickly. If liming a greater soil depth is required, then the recommended values must be increased accordingly.

Sample Questions

pH

1. What is the pH if the [H$^+$] is 0.0005 M?
2. Which solution has more H$^+$, one at pH 7 or one at pH 9?
3. If the [H$^+$] concentration is 0.01 M, what is the pH?
4. Calculate the [H$^+$] at pH 9.
5. Calculate the [H$^+$] at pH 6.5.

Liming

1. What is the CCE of hydrated lime that is only 95 percent pure?
2. What is the ECCE of pure calcitic limestone for which 75 percent passes a 10 mesh and 65 percent passes a 50 mesh?
3. What is the ECCE of dolomitic limestone that is 95 percent pure and is composed of particles that all pass through a 50 mesh screen?
4. If the liming recommendations of a state are based on an ECCE of 80 percent and the soil test suggests that 4 ton acre^{-1} of lime is needed, how much material should you add if you have a source of dolomitic limestone that is 95 percent pure and for which 95 percent of the particles pass a 10 mesh, but only 50 percent pass a 50 mesh sieve?

5. If the ECCE of a liming material is 70 and the lime rate for the state is based on material with an ECCE of 83 percent how should the lime rate be adjusted if you determine that the ECCE of your liming source is 90 percent?

Lime Requirements

1. If the water pH is measured as 6.5, how much lime is required to raise the soil pH to 6.4?
2. If the measured water pH is 5.5 and the buffer pH is also 5.5, how much lime is required to raise the pH to 6.4?
3. If the water pH is 4.9 and the buffer pH is unknown, how much lime is required to raise the soil pH to approximately 6.4?
4. If the final soil pH is desired to be 6.4 and the buffer pH is 6.5, how much lime must be added to accomplish this goal if the water pH is 5.7?

20

Cation and Anion Exchange

OBJECTIVE

In this chapter we look at several important calculations involving ion exchange in soil.

- Estimate cation and anion exchange capacity (CEC and AEC) based on characteristics of the soil constituents.
- Calculate base saturation.
- Determine total dissolved solids (TDS) based on electrical conductivity measurements.
- Estimate sodium adsorption ratio (SAR) and exchangeable sodium ratio (ESR) in salt-affected soils.

Overview

Ion exchange is the interchange of ions in the soil solution with ions adsorbed to the surface of soil clay minerals, organic matter, or other soil colloids. The ions can be cations (positively charged) or anions (negatively charged). Ion exchange gives soils the ability to retain nutrients and water, and then make these available to plants. It also allows soils to retain organic and inorganic contaminants applied either intentionally or by accident. Without the ability to exchange and retain ions with the surrounding soil solution, the role of soil would be greatly diminished.

Cation Exchange

In most soils the net surface charge is negative, so cation exchange is the most common type of exchange reaction. A negatively charged soil colloid will retain enough cation charge to balance the net negative charge on the colloid. Exchange reactions are influenced by the types of cations present on the adsorption sites and in the soil solution because of differences in affinity, or strength of adsorption. Aluminum is held tightly within the exchange complex because of a high positive charge and a relatively small size. A simplified order of strength of adsorption is

$$Al^{3+} > Ca^{2+} > Mg^{2+} > K^+ = NH_4^+ > Na^+$$

Another important consideration is the concentration of cations in the surrounding soil solution. As the concentration of a given cation increases in the soil solution, it will more readily replace cations on the exchange sites. This is the basis of many extraction procedures used in soil science. A high concentration of one cation (1 M K^+, for example) is used to replace another cation (NH_4^+) on the soil colloids.

Exchange capacity is the sum total of exchangeable ions that a soil can adsorb. Similarly, it is also used to describe the ability of a soil to hold ions. The vast majority of the exchange complex is occupied by cations because of the net negative charge on most colloids, so we often focus on the **cation exchange capacity.** There may also be at least a small **anion exchange capacity.** The exchange capacity is normally expressed in units of $cmol_c$ kg^{-1}.

The CEC is controlled by four soil factors—(i) soil texture, or more specifically clay content; (ii) clay type; (iii) organic matter content; and (iv) soil pH. The colloids in the soil are the source of charge, so the greater the content of colloids in the soil, either mineral or organic, the greater the CEC. As an example, consider the data in Table 20–1 from a soil sample collected in the A horizon of a soil from a humid environment.

Note that the values are given in concentration of charge ($cmol_c$ kg^{-1}), this is equivalent to milliequivalents per 100 g of soil, so 10 $cmol_c$ kg^{-1} = 10 meq

---

TABLE 20–1 Cations in the A Horizon of a Humid Soil

Cation	Concentration
Ca^{2+}	10.8 $cmol_c$ kg^{-1}
Mg^{2+}	0.12 $cmol_c$ kg^{-1}
K^+	0.94 $cmol_c$ kg^{-1}
Na^+	0.07 $cmol_c$ kg^{-1}
H^+	3.87 $cmol_c$ kg^{-1}
Al^{3+}	0.0 $cmol_c$ kg^{-1}

positive charge per 100 g. The cation exchange capacity is calculated as merely the sum of all cations.

$$CEC = Ca^{2+} + Mg^{2+} + K^+ + Na^+ + H^+ + Al^{3+}$$ **20-1**

Example 20–1

What is the CEC of the soil sample given the data in Table 20–1?

Solution

For this example we find

$$CEC = 10.8 + 0.12 + 0.94 + 0.07 + 3.87 + 0.00$$

$$= 15.80 \, cmol_c \, kg^{-1} \, (15.8 \, meq/100 \, g)$$

You can calculate the percent saturation with a given cation.

Example 20–2

What is the Ca saturation of the soil sample in Table 20–1?

Solution

The percent Ca saturation is

$$\%Ca \text{ saturation} = \left(\frac{Ca^{2+}}{CEC}\right) \times 100 = \left(\frac{10.80}{15.80}\right) \times 100 = 68.4\%$$

or the percent base cation

$$BCS = \left(\frac{Ca^{2+} + Mg^{2+} + K^+ + Na^+}{CEC}\right) \times 100$$

$$= \left(\frac{10.80 + 0.12 + 0.94 + 0.07}{15.80}\right) \times 100 = 75.55$$

The base cation saturation is important because it conveys information about the ability of a soil to provide these important plant nutrients and the relative acidity of the soil solution. We will address this further when we discuss soil acidity.

We can also calculate an estimate of CEC because of the known relationships between clay content, clay type, organic matter content, and CEC (Table 20–2). Consider the following information about a soil:

Organic matter content = 3.0%
Clay content = 25%
Clay type = smectite

**TABLE 20–2   Cation and Anion Exchange Properties of Various Soil
Constituents**

Soil Component	Approximate Cation Exchange Capacity ($cmol_c$ kg^{-1})
Soil organic matter (humus) (pH dependent)	200
Vermiculite clay	150
Montmorillionite clay	100
Smectite	100
Allophane	30
Chlorite	30
Illite clay	30
Micas	30
Kaolinite clay	8–10
Gibbsite (Al mineral)	4
Goethite (Fe mineral)	4

The contribution of charge from both clay and organic matter (OM) can be calculated. Organic matter has a typical CEC of 200 $cmol_c$ kg^{-1} (Table 20–2).

Example 20–3

What is the CEC contributed by OM in a soil sample with 3.0 percent organic matter (or 0.03 kg organic matter per kilogram of soil)?

Solution

$$\frac{200 \text{ cmol}_c}{\text{kg OM}} \times \frac{0.03 \text{ kg OM}}{\text{kg soil}} = \frac{6 \text{ cmol}_c}{\text{kg soil}}$$

Smectite clay has a typical CEC of 100 $cmol_c$ kg^{-1} (Table 20–2).

Example 20–4

What is the contribution to the CEC of smectite clay in a soil sample with 25 percent smectite clay?

Solution

$$\frac{100 \text{ cmol}_c}{\text{kg clay}} \times \frac{0.25 \text{ kg clay}}{\text{kg soil}} = \frac{25 \text{ cmol}_c}{\text{kg soil}}$$

Example 20–5

What is the total CEC of this soil?

Solution

$$\frac{6 \text{ cmol}_c}{\text{kg soil}} + \frac{25 \text{ cmol}_c}{\text{kg soil}} = \frac{31 \text{ cmol}_c}{\text{kg soil}}$$

Anion Exchange

Less common in most soils are anion exchange reactions. In such instances, a soil colloid with positive exchange sites in the ion exchange complex may retain negatively charged ions. These anions may be replaced by other anions in the soil solution.

Salinity and Sodicity

Several measures are used to evaluate salinity (high salt content) and sodicity (high sodium content) in soil—**total dissolved solids (TDS), electrical conductivity (EC), exchangeable sodium percentage (ESP),** and **sodium adsorption ratio** (SAR).

Total dissolved solids are the mass of solid material dissolved in a solution, usually expressed as mg L^{-1}. Irrigation water may have TDS values ranging from 5 to 1000 mg L^{-1}, while soil extracts can have TDS contents between 1000 and 12,000 mg L^{-1}. Total dissolved solids can be evaluated by evaporating a solution to dryness. But TDS are more often related by empirical equations to electrical conductivity, which is the conductance of electricity through a solution measured in siemans (S) or decisiemans (dS) per meter and can be easily measured with a conductance meter. The more dissolved salts in solution, the greater the electrical conductivity.

$$\Sigma(\text{Cations or anions, meq } L^{-1}) \approx EC \text{ (dS } m^{-1}) \times 10 \qquad \textbf{20-2}$$

$$\text{TDS (mg } L^{-1}) \approx EC \text{ (dS } m^{-1}) \times 640 \qquad \textbf{20-3}$$

(for solutions with mostly Na salts)

$$\text{TDS (mg } L^{-1}) \approx EC \text{ (dS } m^{-1}) \times 800 \qquad \textbf{20-4}$$

(for solutions with mostly Ca salts)

Another useful relationship is that of osmotic potential in soil to electrical conductivity.

$$\text{Osmotic potential } (\psi_\pi)(\text{bars}) \approx \text{EC (dS m}^{-1}) \times (-0.36) \qquad \textbf{20-5}$$

Example 20–6

What is the TDS, EC, and osmotic potential of a solution that contains 1 meq L^{-1} Ca^{2+}, 2 meq L^{-1} Mg^{2+}, and 4 meq L^{-1} Na^+?

Solution

$$\text{The sum of cations} = 1 + 2 + 4 = 7 \text{ meq L}^{-1}$$

$$7 \text{ meq L}^{-1} \div 10 = 0.7 \text{ dS m}^{-1} \text{ for the EC}$$

$$0.7 \text{ dS m}^{-1} \times -0.36 = -0.25 \text{ bars or } -0.025 \text{ Mpa for the osmotic potential}$$

$$0.7 \text{ dS m}^{-1} \times 640 = 448 \text{ mg L}^{-1} \text{ TDS}$$

A saline soil has enough soluble salts to impair growth. The electrical conductivity will be ≥ 4 dS m^{-1} in a saturated soil paste.

Two parameters used to evaluate the potential sodicity of soil are the sodium adsorption ratio and the exchangeable sodium percentage (ESP).

$$\text{SAR} = \frac{[Na^+]}{(0.5\ Ca^{2+} + 0.5\ Mg^{2+})^{1/2}} \qquad \textbf{20-6}$$

$$\text{ESP} = \frac{\text{exchangeable Na}^+, \text{cmol}_c \text{ kg}^{-1}}{\text{cation exchange capacity, cmol}_c \text{ kg}^{-1}} \times 100 \qquad \textbf{20-7}$$

Example 20–7

For a soil that has 10 mmol$_c$ Ca^{2+}, 5 mmol$_c$ Mg^{2+}, and 1 mmol$_c$ Na^+ per kilogram, what is its SAR?

Solution

$$\text{SAR} = \frac{[Na^+]}{(0.5\ Ca^{2+} + 0.5\ Mg^{2+})^{1/2}}$$

$$\text{SAR} = \frac{(1 \text{ mmol}_c \text{ Na}^+)}{[(0.5)(10 \text{ mmol}_c \text{ Ca}^{2+}) + (0.5)(5 \text{ mmol}_c \text{ Mg}^{2+})]^{1/2}}$$

$$\text{SAR} = 0.36$$

TABLE 20–3 Classifications for Salt-Affected Soils

Normal Soils	Saline Soils	Sodic Soils	Saline-Sodic Soils
$EC < 4$ dS m^{-1}	$EC > 4$ dS m^{-1}	$EC < 4$ dS m^{-1}	$EC > 4$ dS m^{-1}
$SAR < 13$	$SAR < 13$	$SAR > 13$	$SAR > 12$
$ESP < 15\%$	$ESP < 15\%$	$ESP > 15\%$	$ESP > 15\%$

Example 20–8

What is the ESP of a soil that has 15 cmol$_c$ kg^{-1} Na$^+$ and a CEC of 150 cmol$_c$ kg^{-1}?

Solution

$$ESP = \frac{\text{exchangeable Na}^+, \text{cmol}_c \text{ kg}^{-1}}{\text{cation exchange capacity, cmol}_c \text{ kg}^{-1}} \times 100$$

$$ESP = \frac{15 \text{ cmol}_c \text{ kg}^{-1} \text{ Na}^+}{150 \text{ cmol}_c \text{ kg}^{-1}} \times 100 = 10\%$$

A sodic soil has 13 to 15 percent of the CEC occupied by Na$^+$. Saline-sodic soils have both an EC ≥ 4 dS m^{-1} and 13 to 15 percent or more of the CEC occupied by Na$^+$, which is approximately equivalent to an SAR of 15 (Table 20–3). The relationship between ESP and SAR is approximately

$$\frac{ESP}{(100 - ESP)} = 0.015 \text{ SAR} \qquad \boxed{\textbf{20-8}}$$

Example 20–9

What is the estimated SAR if the ESP is 7 percent?

Solution

$$\frac{ESP}{(100 - ESP)} = 0.015 \text{ SAR}$$

$$SAR = \frac{[7/(100 - 7)]}{0.015} = 5$$

Sample Problems

Calculating and Estimating CEC

Use the data set below to answer Questions 1 to 4.

Ca^{2+} 7 meq per 100 g soil
Mg^{2+} 6 meq per 100 g soil
K^+ 2 meq per 100 g soil
H^+ 5 meq per 100 g soil
Al^{3+} 3 meq per 100 g soil

1. Convert the above-mentioned quantities of exchangeable ions into $cmol_c\ kg^{-1}$.
2. What is the CEC?
3. What is the percent base saturation?
4. What is the Mg^{2+} saturation?
5. For a soil consisting of 2 percent organic matter and 20 percent montmorillonitic clay, estimate the CEC.

Salinity and Sodicity

1. What is the estimated EC of a solution that has 1 meq L^{-1} Na^+, 3 meq L^{-1} K^+, 3 meq L^{-1} HCO_3^-, and 4 meq L^{-1} Ca^{2+}?
2. If TDS are calculated as 250 mg L^{-1}, what is a likely value for EC in a solution dominated by Ca salts?
3. If the osmotic pressure (ψ_π) of a soil solution is -0.15 Mpa, what is an estimated EC in decisiemans per meter?
4. What is the SAR of a soil that has (per kilogram) 5 mmol Na^+, 15 mmol Ca^{2+}, 5 mmol K^+, and 2 mmol Mg^{2+}?
5. What is the ESP of the soil in Table 20–1?
6. What is the classification of a soil that has an EC of 5 dS m^{-1} and 15 percent exchangeable Na^+?
7. What is the ESP of a soil that has 10, 5, and 3 $cmol_c\ kg^{-1}$ Ca^{2+}, Mg^{2+}, and Na^+, respectively? Is this a sodic soil?

21

Calculating Fertilizer Application Rates and Nutrient Availability

---

OBJECTIVE

In this chapter you will

- review the meaning of fertilizer labels.
- convert between elemental and oxide values for nutrient concentrations.
- calculate fertilizer requirements based on soil test results.
- calculate nutrient delivery and availability from organic amendments.

Overview

Plants require 14 essential nutrients from soil in addition to carbon (C), hydrogen (H), and oxygen (O), which they obtain from air and water (Table 21–1). A few other nutrients such as cobalt (Co), silicon (Si), and sodium (Na) appear to be required by some but not all plants (Tisdale & Nelson, 1975). With the exception of plants able to form symbiotic and asymbiotic associations with N-fixing bacteria to supply their N, these nutrients must come from the mineralization of organic matter in soil, weathering of soil minerals, or nutrients bound to cation and anion exchange sites of the soil. When these processes are insufficient to meet plant needs, fertilization is required. Being comfortable with calculating fertilizer requirements is therefore an important skill.

TABLE 21–1   Required Nutrients in Plant Production

Class	Name	Symbol	Plant Available Forms
Macronutrient	Nitrogen	N	Ammonium (NH_4^+) Nitrate (NO_3^-)
	Phosphorus	P	Orthophosphate ($H_2PO_4^-$, HPO_4^{2-})
	Potassium (Potash)	K	Potassium ion (K^+)
	Calcium	Ca	Divalent Ca (Ca^{2+})
	Magnesium	Mg	Divalent Mg (Mg^{2+})
	Sulfur	S	Sulfate (SO_4^{2-})
Micronutrient	Boron	B	Boric Acid (H_3BO_3) Borate ($H_2BO_3^-$)
	Chlorine	Cl	Chloride ion (Cl^-)
	Copper	Cu	Cupric ion (Cu^{2+})
	Iron	Fe	Ferrous iron (Fe^{2+}) Ferric iron (Fe^{3+})
	Manganese	Mn	Manganous ion (Mn^{2+})
	Molybdenum	Mo	Molybdate ion (MoO_4^{2-})
	Nickel	Ni	Divalent nickel (Ni^{2+})
	Zinc	Zn	Divalent zinc (Zn^{2+})

Fertilizer Conversions

The three most limiting nutrients in soil are typically N, P, and K. Most commercial fertilizers are blended mixes of one or more of these nutrients expressed as %N, %P_2O_5, and %K_2O. To convert from %P_2O_5 to %P on an elemental basis, multiply by 0.44. To convert from %K_2O to %K on an elemental basis, multiply by 0.83. The values 0.44 and 0.88 reflect that 44 percent of the mass of P_2O_5 is composed of elemental P, and that 83 percent of the mass of K_2O is composed of elemental K. Similarly, to convert from CaO and MgO to the elemental forms Ca and Mg, multiply by 0.715 and 0.602, respectively.

Example 21–1

A fertilizer has an analysis listed as 18-46-22. What does this mean?

Solution

This means the fertilizer consists of 18 percent elemental N, 46 percent P as P_2O_5, and 22 percent K as K_2O. The remaining material is filler.

Example 21–2

On an elemental basis, what %P is in a fertilizer listed as 10-34-0?

Solution

$$0.44 \times 34\% = 15\%$$

Example 21–3

On an elemental basis, what %K is in a fertilizer listed as 13-0-44?

Solution

$$0.83 \times 44\% = 36\%$$

The analysis of the fertilizers differs considerably depending on whether they are designed to principally deliver N, P, or K (Table 21–2).

Analytical labs may provide a nutrient analysis or requirement on an elemental basis, in which case you have to convert to the oxide forms of P (P_2O_5)

TABLE 21–2 Typical Analyses of Various Fertilizers

Major Element	Source	Composition	Analysis (%N-%P_2O_5-%K_2O)
N	Anhydrous ammonia	NH_3	82-0-0
	Urea	$(NH_2)_2CO$	46-0-0
	Ammonium nitrate	NH_4NO_3	34-0-0
	Ammonium sulfate	$(NH_4)_2SO_4$	21-0-0
	UAN	Urea + Ammonium nitrate	28-0-0 to 32-0-0
P	Superphosphate	$Ca(H_2PO_4)_2$	0-20-0
	Triple superphosphate (concentrated)	$Ca(H_2PO_4)_2$	0-46-0
	Monoammonium phosphate	$NH_4H_2PO_4$	11-52-0
	Diammonium phosphate	$(NH_4)_2HPO_4$	18-46-0
	Ammonium polyphosphate		10-34-0 to 11-37-0
K	Potassium chloride	KCl	0-0-60
	Potassium sulfate	K_2SO_4	0-0-50
	Potassium nitrate	KNO_3	13-0-44

and K (K_2O). To go from elemental P to the oxide form, simply divide by 0.44 (or multiply by the inverse, 2.27). Similarly, to go from elemental K to K_2O, divide by 0.83 (or multiply by the inverse, 1.2). To go from Ca to CaO, divide by 0.715 (or multiply by 1.39). To go from Mg to MgO, divide by 0.602 (or multiply by 1.66).

$$P_2O_5 \times 0.44 = P \qquad P \times 2.27 = P_2O_5 \qquad \boxed{\text{21-1}}$$

$$K_2O \times 0.83 = K \qquad K \times 1.2 = K_2O \qquad \boxed{\text{21-2}}$$

$$CaO \times 0.715 = Ca \qquad Ca \times 1.39 = CaO \qquad \boxed{\text{21-3}}$$

$$MgO \times 0.602 = Mg \qquad Mg \times 1.66 = MgO \qquad \boxed{\text{21-4}}$$

For minor elements that are usually reported in parts per million, a useful conversion figure to remember is that

$$\text{ppm} \times 0.002 = \text{lb ton}^{-1} \qquad \boxed{\text{21-5}}$$

Example 21–4

If a fertilizer analysis is 10-15-20 on an elemental basis, what is the analysis on a corresponding oxide basis?

Solution

$$10 \div 1 = 10$$

$$15 \div 0.44 = 34$$

$$20 \div 0.83 = 24$$

$$10\text{-}34\text{-}24$$

If you get confused about whether to multiply or divide by the conversion factor, remember that percent analysis reported on an oxide basis will always be greater than that reported on an elemental basis (with the exception of N, which does not change). If it's not, your math's wrong—try reversing your operation.

Calculating Fertilizer Applications

Good soil stewardship for crop production means taking frequent soil samples and replacing nutrients as they are removed from the soil. It is important to be able to utilize soil test data to calculate an appropriate amount of fertilizer to use from the perspective of economics (paying for more than you need) and the environment (reducing the potential for runoff and off-site contamination) perspective.

Example 21–5

If the soil test results on an elemental basis call for 25 lb P per acre and 150 lb K per acre, how much superphosphate (0-20-0) and KCl (0-0-60) would you add?

Solution

The soil test results are on an elemental basis, but most fertilizers are sold on an oxide basis, so a conversion has to be made.

Step 1.

$$25 \text{ lb P} \div 0.44 = 57 \text{ lb P}_2\text{O}_5$$

$$150 \text{ lb K} \div 0.83 = 181 \text{ lb K}_2\text{O}$$

Step 2. Divide the values for P_2O_5 and K_2O by the amount of P_2O5 and K_2O actually present in the fertilizer. 0-20-0 means 20 lb P_2O_5 per 100 lb, while 0-0-60 means 60 lb K_2O per 100 lb.

$$57 \text{ lb} \div 20 \text{ lb P}_2\text{O}_5/100 \text{ lb 0-20-0} = 285 \text{ lb 0-20-0 per acre}$$

$$181 \text{ lb} \div 60 \text{ lb K}_2\text{O}/100 \text{ lb 0-0-60} = 302 \text{ lb 0-0-60 per acre}$$

Some soil testing services have fertilizer calculators on their web sites that enable you or your supplier to calculate an appropriate fertilizer blend to meet a soil test analysis. You can view one example at the University of Kentucky Regulatory Services Soil Testing Lab Web site at <http://www.uky.edu> Search Terms: Ag, Regulatory Services.

Mixing Fertilizers

Unless you're willing to accept premixed fertilizers, you may want to have a custom blend of fertilizers made to meet the specific needs of a field or crop. To make blends, you have to know how much of each carrier to add to a mix so that the final result has the appropriate amounts of each essential nutrient. The formula to use to calculate the appropriate blends is (Plaster, 1997)

$$Y = \frac{A \times B}{C} \qquad \boxed{\textbf{21-6}}$$

where

Y = kilograms (pounds) of carrier for each element
A = kilograms (pounds) of mixed fertilizer needed
B = percentage of the element needed
C = percentage of the element in the carrier

Example 21–6

How would you make 1 Mg (1000 kg) of 10-10-20 from the following carriers:

Ammonium nitrate 33-0-0
Triple superphosphate 0-46-0
Muriate of potash 0-0-60

Solution

Ammonium nitrate

$$Y = \frac{(1000)(10)}{33} = 303 \text{ kg}$$

Triple superphosphate

$$Y = \frac{(1000)(10)}{46} = 217 \text{ kg}$$

Muriate of potash

$$Y = \frac{(1000)(20)}{60} = 333 \text{ kg}$$

$$303 \text{ kg} + 217 \text{ kg} + 333 \text{ kg} = 853 \text{ kg}$$

The remaining mass (147 kg) would be replaced by inert filler.

What may not be obvious from this calculation, but which becomes apparent if you try working with other combinations of nutrients, is that if the desired nutrient concentration is high, but the starting purity of the mix elements is low, then you end up with a blend that has considerably different ratios than you intended.

Example 21–7

How would you make 1 Mg (1000 kg) of 20-20-20 from the following carriers:

UAN 28-0-0
Triple superphosphate 0-46-0
Potassium sulfate 0-0-50

Solution

Ammonium nitrate

$$Y = \frac{(1000)(20)}{28} = 714 \text{ kg}$$

Triple superphosphate

$$Y = \frac{(1000)(20)}{46} = 435 \text{ kg}$$

Sulfate of potash

$$Y = \frac{(1000)(20)}{50} = 400 \text{ kg}$$

$$714 \text{ kg} + 435 \text{ kg} + 400 \text{ kg} = 1549 \text{ kg}$$

$$\text{Analysis} = 13\text{-}13\text{-}13$$

One option is to simply apply the 1549 kg of blended fertilizer uniformly over the intended area. The required amounts of the required nutrients will be added. Another option is to make 1000 kg of 10-10-10, which has exactly the same ratio (1-1-1), and add double the amount.

Available Nutrients in Manures

Animal wastes contribute a sizeable portion of the nutrients supplied to plants in some operations. Animal wastes suffer from two major problems as nutrient sources—(1) they have variable nutrient composition and moisture content and (2) they release their nutrients at variable rates with time. You can see the former in Table 21–3, which reports the analysis for various types of manures on a dry weight basis.

Between 40 and 80 percent of the P and 90 percent of the K is available depending on the type of animal waste and its method of addition to soil. Nitrogen availability is much more variable depending on whether the N is primarily in the form of ammonium (NH_4^+) (in poultry manure), urea, or organic N. Without incorporation, much of the urea and NH_4^+-N can be lost by volatilization within 7 days. As a general rule, 50 to 60 percent of the organic N in animal wastes becomes available in the year of application.

TABLE 21–3   Analysis of Various Animal Manures on an "as-is" Basis

Type	N (lb ton^{-1})	P$_2$O$_5$ (lb ton^{-1})	K$_2$O (lb ton^{-1})	%Moisture
Beef solid	11	7	10	52
Dairy	11	9	12	60
Horse	12	10	12	50
Poultry broiler				
Fresh	55	55	45	20
Stockpiled	40	80	35	20
Cake	60	70	40	30
Pullet	40	68	40	25
Breeder	35	55	30	40
Poultry layer	30	40	30	40
Goat	22	12	24	52
Sheep	21	9	19	52
Rabbit	24	25	11	50

Note. Rasnake, M., Stipes, D., Sikora, F., Duncan, H., & Abnee, A. (2002). *Nutrient management planning guidelines to comply with the Kentucky Agriculture Water Quality Act. ENRI-136.* Lexington, KY: Cooperative Extension Service. University of Kentucky.

Animal wastes also have variable water contents and many testing services report the nutrient analysis on a dry weight basis. It is important to base calculations on nutrient delivery after the proper conversions to a wet weight basis.

Example 21–8

A manure contains four percent nitrogen, two percent phosphorus, two percent potassium, and 80 percent dry matter (this is very dry material as far as animal manures go; 50 percent dry matter content is common). What is the nutrient analysis on an oxide basis per ton of wet material and how much will be available the first year?

Solution

Step 1. Determine moist weight concentrations.

$$4\% \text{ N} \times 0.8 = 3.2\% \text{ N}$$

$$2\% \text{ P} \times 0.8 = 1.6\% \text{ P}$$

$$2\% \text{ K} \times 0.8 = 1.6\% \text{ K}$$

Step 2. Calculate the oxide concentrations.

$$3.2\% \text{ N} \div 1 = 3.2\% \text{ N}$$

$$1.6\% \text{ P} \div 0.44 = 3.6\% \text{ P}_2\text{O}_5$$

$$1.6\% \text{ K} \div 0.83 = 1.9\% \text{ K}_2\text{O}$$

Step 3. Calculate the amount of each nutrient available that year.

$$3.2\% \text{ N} \times 0.6 = 1.9\% \text{ N}$$

$$3.6\% \text{ P}_2\text{O}_5 \times 0.8 = 2.9\% \text{ P}_2\text{O}_5$$

$$1.9\% \text{ K}_2\text{O} \times 0.9 = 1.7\% \text{ K}_2\text{O}$$

Example 21–9

If a soil test calls for 150 lb N, 75 lb P_2O_5, and 125 lb K_2O, how much poultry manure should be added, assuming it has 33 percent moisture and a dry weight analysis of 3% N, 3% P_2O_5, and 2% K_2O, and that it is added strictly to meet the N needs of the growing crop? Assume no volatilization loss occurs and there is 60 percent availability of all nutrients in the year of application.

Solution

Step 1. Convert the nutrient analyses to a wet weight basis. Only 67 percent (0.67) of the material is solid.

$$4\% \text{ N} \times 0.67 = 2.7\% \text{ N}$$

$$3.0\% \text{ P}_2\text{O}_5 \times 0.67 = 2.0\% \text{ P}_2\text{O}_5$$

$$2\% \text{ K}_2\text{O} \times 0.67 = 1.3\% \text{ K}_2\text{O}$$

Step 2. Calculate the availability in the year of application.

$$2.7\% \text{ N} \times 0.6 = 1.62\% \text{ N}$$

$$2.0\% \text{ P}_2\text{O}_5 \times 0.6 = 1.2\% \text{ P}_2\text{O}_5$$

$$1.3\% \text{ K}_2\text{O} \times 0.6 = 0.8\% \text{ K}_2\text{O}$$

Step 3. Calculate the nutrient content per wet ton of manure.

$$2000 \text{ lb ton}^{-1} \times 1.62\% \text{ N} = 32.4 \text{ lb N/ton}$$

$$2000 \text{ lb ton}^{-1} \times 1.2\% \text{ P}_2\text{O}_5 = 24 \text{ lb P}_2\text{O}_5/\text{ton}$$

$$2000 \text{ lb ton}^{-1} \times 0.8\% \text{ K}_2\text{O} = 16 \text{ lb K}_2\text{O}/\text{ton}$$

Step 4. Divide the required nutrient by the nutrient content of the manure.

$$150 \text{ lb N} \div 32.4 \text{ lb N/ton} = 4.6 \text{ ton}$$

Example 21–9 illustrates one of the ongoing problems with adding animal wastes to meet fertilizer needs on cropland. Because most wastes have excess P relative to N for a plant's needs, applications based on N tend to over apply P, leading to potential environmental problems. In Example 21–9, for instance, the application rate of 4.6 ton acre^{-1} will deliver 110 lb available P$_2$O$_5$, which is 47 percent more P than is required. Over the long term this can lead to excessive P buildup in soil.

Many state extension services also have manure calculators that will allow you to calculate the amount of manure necessary to meet a particular soil test recommendation based on typical nutrient contents in various manures. You can view an example of a manure calculator at the University of Kentucky Regulatory Services Soil Testing Lab Web site at <http://www.uky.edu> Search Terms: Ag, Regulatory Services.

Liquid Manures

One feature of some **confined animal operations** (CAFOs) is that the manure is stored as part of its pretreatment in aerobic or anaerobic lagoons. These lagoons are periodically agitated and surface applied or injected into cropland. Application rates of liquid manures are usually reported as pounds per gallon (or 100 gal) or kilograms per cubic meter (there are 1000 L in 1 m^3).

The nutrient content of liquid manures varies considerably, but a typical liquid dairy manure might contain (in 1000 gal) 22.2 lb total N, 9.2 lb ammonium N (NH$_4$-N), 13 lb P$_2$O$_5$, and 25 lb K$_2$O (NRCS, 2001).

The biggest issue in working with liquid manure rates (that is, after assessing the actual nutrient content) is being able to work with volumetric waste applications. There are some useful conversions to know.

To convert pounds per ton to pounds per 1000 gal

$$\text{lb ton}^{-1} \times 4.17 = \text{lb}/1000 \text{ gal}$$ **21-7**

To convert pounds per 1000 gal to pounds per ton

$$\text{lb}/1000 \text{ gal} \times 0.2398 = \text{lb ton}^{-1}$$ **21-8**

It is also useful to know that

$$^1/_2 \text{ acre inch irrigation} = 13,600 \text{ gal acre}^{-1}$$

$$1 \text{ acre inch irrigation} = 27,000 \text{ gal acre}^{-1}$$

$$1 \text{ gal water} = 8.34 \text{ lb}$$

Example 21–10

A farmer has a soil test requirement for 160 lb N per acre. How many gallons of liquid dairy manure will he need to apply if his manure analysis indicates that 20 lb N, 10 lb P$_2$O$_5$, and 20 lb K$_2$O are present per 1000 gal? How much additional P and K will he be adding with that waste?

Solution

Step 1. Assess the nutrient content of the liquid dairy manure.

From the information above it is given that

$$N = 20 \text{ lb}/1000 \text{ gal}$$

$$P_2O_5 = 9 \text{ lb}/1000 \text{ gal}$$

$$K_2O = 15 \text{ lb}/1000 \text{ gal}$$

Step 2. Determine the desired application rates.

$$160 \text{ lb N} \times \frac{1000 \text{ gal}}{20 \text{ lb N}} = 8000 \text{ gal}$$

Step 3. How much additional P and K are added?

$$8000 \text{ gal} \times \frac{9 \text{ lb P}_2\text{O}_5}{1000 \text{ gal}} = 72 \text{ lb P}_2\text{O}_5$$

$$8000 \text{ gal} \times \frac{15 \text{ lb K}_2\text{O}}{1000 \text{ gal}} = 120 \text{ lb K}_2\text{O}$$

Example 21–11

If only 60 percent of the N applied in Example 21–10 was available, how much more liquid manure will the farmer need to apply?

Solution

$$\frac{20 \text{ lb N}}{1000 \text{ gal}} \times 0.6 \,(60\% \text{ availability}) = 12 \text{ lb N}/1000 \text{ gal}$$

$$160 \text{ lb N} \times \frac{1000 \text{ gal}}{12 \text{ lb N}} = 13{,}300 \text{ gal}$$

The farmer will have to apply $13{,}300 - 8000 = 5300$ gal more to meet the N requirement.

References

NRCS-590. (2001). Natural resources conservation service conservation practices standard—Nutrient management—Code 590. NRCS, KY. 5-24-01.

Plaster, E. J. (1997). *Soil science & management* (3rd ed., p. 402). Clifton Park, NY: Thomson Delmar Learning.

Tisdale, S. L., & Nelson, W. L. (1975). *Soil fertility and fertilizers* (3rd ed., p. 694). New York: MacMillan Publishing.

Sample Problems

Fertilizer Conversions

1. How much P_2O_5 is in 100 lb of 0-30-15 fertilizer reported on an oxide basis?
2. How much N is in 50 lb of ammonium nitrate (34-0-0)?
3. How much P_2O_5 and K_2O are in 13-13-13 fertilizer?
4. How much N is in ammonium sulfate (21-0-0)?
5. What would be the likely fertilizer label of a fertilizer that contained only urea?
6. What would be the most likely fertilizer label of a package that contained only $Ca(H_2PO_4)_2$?
7. If a fertilizer is 15-15-15 on an oxide basis, what would be its elemental analysis?
8. If a fertilizer is 0-20-40 on an oxide basis, what is its elemental analysis?
9. Is a 30-20-60 fertilizer label realistic?
10. What is the analysis, on an oxide basis, of a 15-10-30 fertilizer labeled on an elemental basis?
11. What is the analysis, on an oxide basis, of a 20-0-15 fertilizer labeled on an elemental basis?

Calculating Fertilizer Applications

Base your answers for Questions 1 and 2 on a fertilizer label of 20-10-15.

1. How many pounds of fertilizer should be added if the fertilizer recommendation is 50 lb P_2O_5 per acre?
2. How many pounds of fertilizer should be added if the fertilizer recommendation is 75 lb K_2O per acre?

Base your calculations for Questions 3 to 5 on a soil test result that calls for 150 lb N, 22 lb P, and 125 lb K per acre.

3. How many pounds of 10-25-10 fertilizer should be applied to meet the N requirement?
4. By how many pounds will the producer exceed the P recommendation if he applies the calculated rate of 10-25-10 fertilizer in the question above?
5. Will an excessive amount of K_2O be applied if the calculated rates from Question 3 are applied?

Mixing Fertilizers

1. How much of each carrier would you have to add to make 1 ton (2000 lb) of a 5-15-5 fertilizer blend if the starting materials were urea (46-0-0), superphosphate (0-20-0), and muriate of potash (0-0-60)?
2. How much of each carrier would you have to add to make 1 ton (2000 lb) of a 5-15-5 fertilizer blend if the starting materials were urea (46-0-0), triple superphosphate (0-46-0), and potassium sulfate (0-0-50)?
3. How much of each carrier would you have to add to make 1 Mg (1000 kg) of a 5-15-30 fertilizer blend given the starting ingredients of UAN (32-0-0), superphosphate (0-20-0), and muriate of potash (0-0-60). If the final mass exceeds 1000 kg, how would you adjust the initial ratio to stay below the 1000 kg limit?
4. If the soil test for a farm calls for 300 lb N, 100 lb P_2O_5, and 150 lb K_2O, how much of each carrier would you have to add to make a blended mix of each element with starting materials of ammonium nitrate (34-0-0), triple super phosphate (0-46-0), and potassium sulfate (0-0-50)?

Available Nutrients in Manures

Use Table 21–3 (on page 182) to help answer the following questions:

1. If the dry weight analysis of dairy solids is 2.4% N, 2% P_2O_5, 2% K_2O, and the manure is 80 percent liquid, how much of each nutrient is present per wet ton?
2. How much elemental N, P, and K will composted broiler litter provide if the dried material has an analysis of 3.3% N, 5.5% P_2O_5, 5.4% K_2O, and 80% solids and is 50 percent available?

3. In Question 2, if the soil test recommendation calls for 100 lb P per acre, how much composted broiler litter should be added and how much additional N will be required if the soil test calls for 100 lb N per acre?

4. Poultry litter has a nutrient analysis of 60 lb N, 55 lb P_2O_5, and 35 lb K_2O per ton of dry litter at 80 percent solids content. If 70 percent of the N is ammonium-N, which can volatilize away as shown in the table below, how much N has the producer probably applied if six days elapse between litter addition and incorporation, and 4 ton of litter was added per acre?

Time	Ammonium-N lost without incorporation
0–2 days	20%
3–4 days	40%
5–6 days	60%
7+ days	80%

5. In Question 4 above, how much N will actually be available to crops the first year?

Liquid Manures

Use Table 21–4 to help you answer some of these questions.

1. A swine producer has a gestation barn that has 60,000 gal of manure ready to be land applied to a cornfield. The producer plans to inject the manure in the spring, which will prevent runoff and volatilization of nutrients. An analysis of the manure indicates that there is 20 lb of total nitrogen per 1000 gal of manure and 5 lb of total phosphorus (TP) per 1000 gal. Availability indices estimate that 60 percent of the total nitrogen (TN) and 80 percent of the total phosphorus will be available for crop production. Calculate the manure application rate in gallons per acre (to the nearest 50 gal) if the producer wants to apply 150 lb of N per acre? How much TP (in pounds per acre) will be applied using the recommended N application rate?

2. A swine producer has a finishing barn that has 120,000 gal of manure ready to be land applied to a cornfield. The producer plans to inject the manure in the spring, which will prevent runoff and volatilization of nutrients. An analysis of the manure indicates that there is 75 lb of total nitrogen per 1000 gal of manure and 22 lb of total phosphorus per 1000 gal. Availability indices estimate that 60 percent of the total nitrogen and 80 percent of the total phosphorus will be available for crop production. Soil fertility recommendations suggest applying 170 lb of N per acre and 30 lb of P per acre. Calculate the manure application rate in gallons per acre (to the nearest 50 gal), to meet the N requirement, and calculate the amount of P applied per acre.

TABLE 21–4 Nutrients Removed by Crops

Crop	Yield Unit	lb per Yield Unit	Nutrients Removed per Yield Unit		
			N	P_2O_5	K_2O
Silage					
Corn	ton	2000	7.5	3.6	8.0
Small grain	ton	2000	11.7	5.0	7.6
Hay					
Alfalfa hay	ton	2000	50.0	14.0	55.0
All other grass/ legume	ton	2000	35.0	12.0	53.0
Bermuda grass	ton	2000	38.0	9.0	34.0
Eastern gamagrass	ton	2000	35.0	16.0	31.0
Reed canary grass	ton	2000	27.0	8.0	25.0
Warm season Native grass	ton	2000	20.0	6.8	25.0
Grain					
Corn	bushel	56	0.7	0.4	0.35
Barley	bushel	48	0.9	0.4	0.3
Oats	bushel	32	0.6	0.25	0.2
Rye	bushel	56	1.2	0.3	0.3
Sorghum	bushel	56	0.95	0.4	0.3
Soybeans	bushel	60	3.0	0.7	1.1
Winter wheat	bushel	60	1.2	0.5	0.3
Pasture					
Bermuda grass	ton	2000	12.0	3.0	11.0
Grass/legume	ton	2000	11.0	4.0	16.0
Tobacco					
Burley	pound	1	0.07	0.01	0.075
Dark air-cured	pound	1	0.07	0.01	0.06
Dark fire-cured	pound	1	0.07	0.01	0.06

Note. Rasnake, M., Stipes, D., Sikora, F., Duncan, H., & Abnee, A. (2002). *Nutrient management planning guidelines to comply with the Kentucky Agriculture Water Quality Act. ENRI-136.* Lexington, KY: Cooperative Extension Service. University of Kentucky.

Questions 3 to 6 are a little more complicated and challenging—but so are real-world problems.

3. A swine producer has an earthen manure storage structure that has 1,000,000 gal of swine manure that needs to be land applied to a cornfield, before the manure overflows and runs into a blue-line (continuously flowing) stream. The manure contains 24 lb and 10 lb of TN and TP, respectively, per 1000 gal. Availability of TN and TP after being surface applied is 45 and 80 percent, respectively. The producer has a 50-acre cornfield that should yield 150 bushels of corn and requires 150 lb of N per acre. However, repeated applications of manure to this field has increased the soil test phosphorus to a level in which manure application rates are restricted to phosphorus removal rates by the corn (0.15 lb bushel^{-1}). Calculate the manure application rate to meet phosphorus removal. How many additional pounds of N per acre will be needed? How many gallons will be removed from the earthen storage structure for land application?

4. A swine producer plans to apply a manure source with 27 lb of TN and 8 lb of TP per 1000 gal. The producer uses an irrigation gun to apply 27,500 gal per acre to the soil surface, which means that only 45 percent of the N and 80 percent of the P will be available to the crop. How many pounds per acre of N and P are being applied per acre?

5. A dairy farm has 1,000,000 gal of manure with TN and TP concentrations of 24, and 10 lb of TN and TP per 1000 gal. The producer plans to surface apply the manure, and then incorporate within 4 days after application. Therefore, the available N and P should be 55 and 80 percent of the total, respectively. This producer is growing corn after soybeans in a 100-acre field. The producer believes that he needs to apply 0.7 lb of N per bushel of corn produced, and expects a realistic yield of 210 bushels acre^{-1}. The previous soybean crop produced 66 bushels acre^{-1}. If the producer allows a nitrogen credit of 0.5 lb of N per bushel of soybeans, what should his liquid manure application rate be to meet the crop N requirement? Soil test results indicate the available P in the field is 328 lb to the acre. The additional manure P will increase the soil test P at a linear rate. Therefore, every pound of manure TP will increase the soil test P by 1 lb. Using the calculated application rate to meet N requirements, what will the soil test P be at the end of the year?

6. A swine producer buys a used manure application tanker. He loads the tanker with swine manure containing 75 lb of TN per 1000 gal and 18 lb of TP per 1000 gal. The rear axles weigh 32,393 lb and the front hitch weighs 418 lb after loading. The producer applies the load to a cornfield and determines he has applied manure to an area 12 ft wide and 887 ft long. The producer reweighs the application tanker and determines the rear axle weighs 1211 and the front axle weighs 289 lb. Assuming a gallon of manure weighs 8.5 lb, what is the capacity of the tanker (round to the nearest 10 gal)? How much manure was applied per acre (round to the nearest 100 gal)? How much total N? How much total P?

22

Potential Erosion— Working with the Universal Soil Loss Equation

---

OBJECTIVE

Many states use the **universal soil loss equation** (USLE) and its computer-driven version, the **revised universal soil loss equation** (RUSLE), to predict sheet and rill erosion by water. In this chapter you will practice

- determining individual components of the USLE for specific locations.
- calculating the erodibility index (EI) of land.
- calculating soil loss using the USLE.

Overview

The USLE and RUSLE are used to predict sheet and rill erosion by water. For states with significant wind erosion, the **wind erosion equation** (WEQ) is used. The USLE is composed of several factors

$$R \times K \times \mathrm{LS} \times C \times P = \text{tons of soil loss per acre per year (T)}$$

The USLE and RUSLE are fundamental equations used to categorize land as erodible, which has significant regulatory implications for land use and management. Many features of these equations for predicting soil erosion are now in computerized format. This chapter will address the more basic aspects of these equations so that in using more advanced techniques you will have a better appreciation for the source and interpretation of your results.

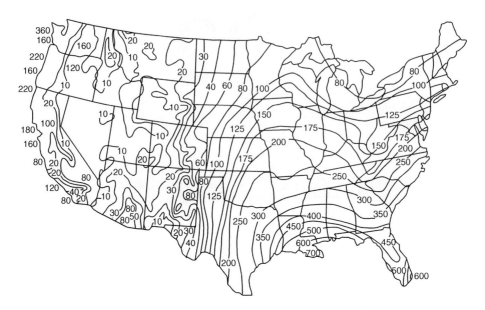

FIGURE 22–1 Isopleths of R values used in the continental USA to predict the erosive effects of rainfall. (*Source:* USDA-NRCS <http://www.nrcs.usda.gov.>.)

Components of the USLE

The **R factor** in the USLE reflects the impact of rainfall on soil surfaces. It depends on the amount, energy, and intensity of rainfall, which is measured at climate stations across each state (Figure 22–1). Specific values for R can be found at the local NRCS office.

Example 22–1

What is the closest approximation of the R factor for southwest Michigan? North Dakota?

Solution

In Figure 22–1 find the line for R passing through southwest Michigan. This has a value of 100. North Dakota, on the other hand, has R values that increase from 30 to 80 as you pass from west to east across the state. In other words, the erosive force of rainfall is higher in southwest Michigan than North Dakota, and increases in North Dakota from west to east.

The **K factor** is assigned to each soil mapping unit. It reflects the influence of soil structure and texture on the erodibility of soil. The K values are reported in county soil surveys for each mapping unit (Figure 22–2).

(The symbol < means less than; > means more than. Entries under "Erosion factors--T" apply to the entire profile. Entries under "Organic matter" apply only to the surface layer. Absence of an entry indicates that data were not available or were not estimated)

Soil name and map symbol	Depth	Clay	Moist bulk density	Permeability	Available water capacity	Soil reaction	Shrink-swell potential	Erosion factors		Organic matter
								K	T	
	In	Pct	g/cc	In/hr	In/in	pH				Pct
BaB----------------	0–6	12–27	1.20–1.40	0.6–2.0	0.18–0.23	4.5–7.3	Low---------------	0.43	3	.5–4
Beasley	6–36	40–60	1.30–1.55	0.2–0.6	0.12–0.18	4.5–7.3	Moderate--------	0.28		
	36–54	40–60	1.50–1.70	0.2–0.6	0.10–0.16	6.6–8.4	Moderate--------	0.28		
	54–60	---	---	0.0–0.6	---	---	-------------------	----		
BcC2---------------	0–5	27–40	1.20–1.40	0.6–2.0	0.14–0.23	4.5–7.3	Low---------------	0.32	3	.5–2
Beasley	5–36	40–60	1.30–1.55	0.2–0.6	0.12–0.18	4.5–7.3	Moderate--------	0.28		
	36–54	40–60	1.50–1.70	0.2–0.6	0.10–0.16	6.6–8.4	Moderate--------	0.28		
	54–60	---	---	0.0–0.6	---	---	-------------------	----		
BeC3, BeD3------	0–4	40–60	1.20–1.40	0.6–2.0	0.14–0.23	4.5–7.3	Low---------------	0.32	3	.5–2
Beasley	4–36	40–60	1.30–1.55	0.2–0.6	0.12–0.18	4.5–7.3	Moderate--------	0.28		
	36–54	40–60	1.50–1.70	0.2–0.6	0.10–0.16	6.6–8.4	Moderate--------	0.28		
	54–60	---	---	0.0–0.6	---	---	-------------------	----		
CaB, CaC, CaD--	0–8	12–27	1.20–1.40	2.0–6.0	0.16–0.22	4.5–6.5	Low---------------	0.28	4	1–4
Carpenter	8–42	18–35	1.20–1.50	0.6–2.0	0.10–0.20	4.5–6.5	Low---------------	0.28		
	42–60	30–55	1.20–1.60	0.06–0.6	0.07–0.16	4.5–6.0	Moderate--------	0.28		
	60–70	---	---		---	---	-------------------	----		
CbF2:	0–6	12–27	1.20–1.40	2.0–6.0	0.16–0.22	4.5–6.5	Low---------------	0.28	4	1–4
Carpenter-------	6–42	18–35	1.20–1.50	0.6–2.0	0.10–0.20	4.5–6.5	Low---------------	0.28		
	42–60	30–55	1.20–1.60	0.06–0.6	0.07–0.16	4.5–6.0	Moderate--------	0.28		
	60–70	---	---		---	---	-------------------	----		
Lenberg---------	0–4	12–27	1.30–1.50	0.6–2.0	0.18–0.23	4.5–7.3	Low---------------	0.43	3	.5–3
	4–14	35–60	1.40–1.60	0.2–0.6	0.10–0.19	4.5–5.5	Moderate--------	0.37		
	14–24	40–60	1.40–1.65	0.2–0.6	0.10–0.18	4.5–5.5	Moderate--------	0.37		
	24–38	40–60	1.40–1.65	0.2–0.6	0.10–0.16	4.5–5.5	Moderate--------	0.28		
	38–45	---	---		---	---	-------------------	----		
CeB----------------	0–9	10–27	1.20–1.40	2.0–6.0	0.16–0.22	5.1–7.3	Low---------------	0.28	4	1–4
Chenault	9–42	18–35	1.20–1.50	0.6–2.0	0.10–0.20	5.1–6.5	Low---------------	0.28		
	42–65	40–55	1.30–1.60	0.6–2.0	0.07–0.16	5.6–7.3	Moderate--------	0.28		
ChC2, ChD2------	0–11	12–27	1.20–1.40	2.0–6.0	0.11–0.18	3.6–6.5	Low---------------	0.37	3	1–3
Christian	11–20	25–42	1.20–1.50	0.6–2.0	0.14–0.22	3.6–6.5	Moderate--------	0.28		
	20–67	40–60	1.30–1.60	0.6–2.0	0.10–0.16	3.6–6.5	Moderate--------	0.28		
	67–75	40–60	1.30–1.60	0.6–2.0	0.10–0.16	3.6–6.5	Moderate--------	0.28		
CnD3--------------	0–11	27–40	1.20–1.50	0.6–2.0	0.14–0.22	3.6–6.5	Low---------------	0.37	3	<2
Christian	11–20	25–42	1.20–1.50	0.6–2.0	0.14–0.22	3.6–6.5	Moderate--------	0.28		
	20–67	40–60	1.30–1.60	0.6–2.0	0.10–0.16	3.6–6.5	Moderate--------	0.28		
	67–75	40–60	1.30–1.60	0.6–2.0	0.10–0.16	3.6–6.5	Moderate--------	0.28		
CrB----------------	0–10	15–27	1.20–1.40	0.6–2.0	0.19–0.23	5.1–7.3	Low---------------	0.32	5	2–4
Crider	10–28	18–35	1.20–1.45	0.6–2.0	0.18–0.23	5.1–7.3	Low---------------	0.28		
	28–62	30–60	1.20–1.55	0.6–2.0	0.12–0.18	4.5–6.5	Moderate--------	0.28		
CrC2---------------	0–7	15–27	1.20–1.40	0.6–2.0	0.19–0.23	5.1–7.3	Low---------------	0.32	5	2–4
Crider	7–28	18–35	1.20–1.45	0.6–2.0	0.18–0.23	5.1–7.3	Low---------------	0.28		
	28–62	30–60	1.20–1.55	0.6–2.0	0.12–0.18	4.5–6.5	Moderate--------	0.28		

FIGURE 22–2 Soil erosion information is provided in county soil surveys. In this example of physical properties for various soil series in Marion Co. Kentucky, both the *K* and *T* values are provided. (*Source:* USDA-NRCS <http://www.nrcs.usda.gov.>.)

Example 22–2

What is the *K* value assigned to surface layer of Crider soils from Marion Co. Kentucky (Soil Conservation Service, 1991)?

Solution

Part of the soil survey from Marion Co. Kentucky is displayed in Figure 22–2. Crider soils with 2 to 6 percent slopes (CrB) have a *K* value of 0.32 assigned to the upper 10 in of soil.

The **LS factor** is a topographic factor and is due to two components that influence soil erosion—the length of a slope (L) and the steepness of a slope (S). These two factors are typically combined into tables or graphs for convenience (Table 22–1).

TABLE 22–1   Table of LS Values

	Slope Length (ft)										
Slope (%)	20	30	40	50	60	80	100	200	400	600	800	1000
0.5								0.11	0.125	0.135	0.145	0.150
1.0				0.105	0.110	0.120	0.130	0.16	0.195	0.220	0.240	0.260
2.0	0.120	0.140	0.150	0.160	0.170	0.190	0.200	0.240	0.300	0.340	0.380	0.400
3.0	0.180	0.200	0.220	0.230	0.240	0.270	0.280	0.350	0.430	0.490	0.525	0.575
4.0	0.210	0.240	0.280	0.300	0.320	0.360	0.400	0.525	0.700	0.825	0.925	1.000
5.0	0.240	0.290	0.340	0.380	0.420	0.480	0.525	0.750	1.100	1.300	1.500	1.700
6.0	0.300	0.370	0.420	0.470	0.525	0.600	0.675	0.950	1.300	1.650	1.900	2.100
8.0	0.440	0.550	0.625	0.700	0.750	0.900	1.000	1.400	2.000	2.400	2.800	3.200
10.0	0.600	0.750	0.850	0.950	1.050	1.200	1.400	1.900	2.700	3.100	3.800	4.200
12.0	0.800	1.000	1.150	1.250	1.400	1.600	1.800	2.500	3.600	4.400	5.000	5.750
14.0	1.000	1.250	1.400	1.600	1.750	2.000	2.300	3.200	4.260	5.500	6.500	7.000
16.0	1.250	1.550	1.800	2.000	2.200	2.500	2.800	4.000	5.750	7.000	8.000	9.000
20.0	1.800	2.200	2.600	3.200	2.600	3.800	4.000	5.750	8.000	10.00	11.50	13.00
25.0	2.600	3.200	3.700	4.100	4.500	5.250	5.750	8.250	11.50	14.00	16.50	18.50
30.0	3.600	4.400	5.000	6.250	7.000	8.000	11.00	14.00	16.00	19.50		
40.0	5.500	7.000	8.000	9.000	9.750	11.50	12.50	18.00				
50.0	8.000	9.750	11.50	14.00	18.00							

Example 22–3

What is the LS value for an 80 ft long field that has a 12 percent slope associated with it?

Solution

The answer is in the body of Table 22–1. To obtain the LS value read down from the given length (80 ft) and across from the given slope percent (12). The intersection of this column and row (1.600) is the LS value for this particular combination of topographical features.

The **C factor** is a cover management factor that is specific to each region, type of vegetation and soil management. Values are obtained from local NRCS offices. The C values range from lows of 0.001, which may occur in a dense stand of sod to 1.0, which would be bare soil.

The **P factor** is a support practice factor that reflects cultural management that would reduce erosion if used. If no support practices are used, the value is 1.0. Practices such as contour plowing, strip cropping, and terracing all decrease the P value. This is also information that is available from the local NRCS office.

Erodibility Index

The **erodibility index** is a tool to help classify land in terms of its susceptibility to erosion. If soils have an EI greater than 8, they are considered highly erodible and the producer needs to seriously consider their suitability for cultivation. Soil mapping units are likewise considered to be highly erodible if they have an EI greater than 8.

$$EI = \frac{(R \times K \times LS)}{T}$$

22-1

where all terms are previously defined in the USLE.

For the purposes of being included into the conservation reserve program (CRP), some other conditions apply.

- One-third or more of the field acreage has an EI greater than 8.
- The weighted EI for the field is greater than 8.
- At least 50 acres of the field are classified as being highly erodible.

To calculate the weighted field EI, do the following:

- Multiply each individual mapping unit (acres) by the appropriate EI for that mapping unit.
- Add the results together for the entire field.
- Divide by the total field acres.

Example 22–4

Hart Co. Kentucky has an R of 200. A farmer's field has a K value of 0.32 and a T factor of 4 ton $acre^{-1}$ $year^{-1}$. If the predominant land slope is 6 percent and the average field is 100 ft long, what is the erodibility index?

Solution

$$EI = \frac{(R \times K \times LS)}{T} = \frac{(200 \times 0.32 \times 0.675)}{4} = 10.8$$

Based on this result, the field is considered highly erodible. The farmer is potentially losing soil at more than twice the rate that it can tolerate.

Example 22–5

What is the weighted field EI for a farmer's field that consists of 50 acres of Beasley silt loam soils with 2 percent slopes (BaB), 100 acres of Beasley silty clay soils with 6 percent slopes (BcC2), and 20 acres of Beasley silty clay soils with 12 percent slopes (BeC3)? Assume the R factor is 175, the average field length for all three fields is 100 ft, and the K value for each soil is 0.43, 0.32, and 0.32, respectively.

Solution

Step 1. Calculate the EI for each soil using the data contained in Figure 22–2.

$$BaB: \quad \frac{(175 \times 0.2 \times 0.43)}{3} = 5.0$$

$$BcC2: \quad \frac{(175 \times 0.32 \times 0.675)}{3} = 12.6$$

$$BeC3: \quad \frac{(175 \times 0.32 \times 1.80)}{3} = 33.6$$

Step 2. Multiply the EI for each soil by the acreage in each mapping unit, sum, and divide by the total acreage.

$$(5.0) \times (50\ acres) + (12.6) \times (100\ acres) + (33.6) \times (20\ acres)$$

$$= \frac{2182}{170}\ acres = 12.8$$

Based on these results, much more than one-third the field acreage consists of highly erodible soil and the weighted EI exceeds 8. This field is a good candidate for the CRP.

Predicting Soil Loss

With all the necessary information from the local NRCS office and the soil survey for a particular location, it becomes relatively easy to predict soil loss simply by plugging numbers into the USLE. These factors are also available in the RUSLE software. The critical feature, of course, is to determine whether current cultural practices cause the predicted soil loss to exceed the designated T value for each soil mapping unit. The R and K values are essentially fixed for a particular site (i.e., you can't change the weather or soil for a particular field). The LS values, which address topography, are also essentially fixed unless major land reconstruction is used. Consequently, if the T value is exceeded, then the producer must change either cover management or cultural practice, or both, to reduce the extent of soil erosion.

Example 22–6

A farmer in central Kentucky is producing continuous corn on 5 percent slopes by planting up and down slopes and without using any cover management. The K value for the soil is given as 0.40 and the average field length is 50 ft. What is his predicted annual soil loss?

Solution

The R value based on Figure 22–1 (page 192) is probably close to 175. The LS value from Table 22–1 (page 194) is 0.38. With no cover management and plowing up and down slopes, the C and P values are both 1.0 (indicating no benefit of management on soil loss). The predicted soil loss is therefore

$$\text{Soil loss} = R \times K \times \text{LS} \times C \times P = (175) \times (0.40) \times (0.38) \times (1) \times (1)$$

$$= 26.6 \text{ ton acre}^{-1} \text{ year}^{-1}$$

Let's put Example 22–6 in perspective. 26.6 ton acre^{-1} year^{-1} is 53,200 lb acre^{-1} year^{-1}. The top 6 in of topsoil (the plow layer) contains about 2,000,000 lb acre^{-1}. Because of his management practices, this farmer is losing about 3 percent of his topsoil every year through erosion. How long do you think he'll be able to keep farming this way?

Reference

Soil Conservation Service. (1991). *Soil survey of Marion County, Kentucky.* Washington, DC: United States Government Printing Office.

Sample Problems

1. What is the average K value for the Lenberg soil in Marion Co. Kentucky in the first 14 in of soil?

2. What is the LS value for a 20-ft-long field with an average slope of 10 percent?
3. What does the slope of a 50-ft field have to be for it to have an LS less than 1.0?
4. What is the predicted soil loss for a field in continuous corn production with conventional tillage across slope if the field has a 6 percent slope with a length of 100 ft? You can assume that the $R = 200$, $C = 0.345$, and $P = 0.75$.
5. The producer in Question 4 decides to switch to no-tillage as a conservation practice, which alters the C value to 0.070. Does this change the predicted soil loss, and if so, by how much?
6. A field on a Marshall silt loam in Iowa has an average slope of 4 percent, field length of 150 ft, R factor of 150, K factor of 0.37, and T value of 3 ton acre^{-1} year^{-1}. What is its erodibility index?
7. What is the weighted erodibility index for a 150 acre field that has an R of 150, and consists of 100 acres of Carpenter gravelly silt loam ($K = 0.28$, $T = 4$ ton acre^{-1} year^{-1}) with 50 ft slopes and 3 percent grades, and 50 acres of Lenberg silt loam ($K = 0.43$, $T = 3$ ton acre^{-1} year^{-1}) with 25 ft slopes that average 12 percent?
8. Determine the potential soil loss for soil in the Krum series near Ft. Hood, TX, if the R is approximately 270, the K is 0.19, the LS averages 0.69, the C value is 0.1, and no erosion practices are used.
9. What is the predicted soil loss for a Saybrook soil in Illinois that has an R of 175, K of 0.32, 300-ft-long slopes with 2 percent grades, no conservation practices, and a cover crop management which gives a C value of 0.5? If the T value is 5 ton acre^{-1} year^{-1}, does the predicted soil erosion loss exceed this value? Is it highly erodible?
10. A Tama silt loam with a K value of 0.32 is being farmed on 2 percent slopes that average 75 ft in length. The R value is 150 for this area, and the farmer is practicing a row crop-oat-meadow rotation to provide a C value of 0.11. His conservation practice is strictly contour plowing, which gives him a P value of 0.7. The T value for this soil is 5 ton acre^{-1} year^{-1}. Does the farmer need to adjust his management to fall below the acceptable T value for his farm? If so, would it be enough for him to simply add a second year of meadow to his rotation to bring his C value to 0.08, or will he have to use other conservation practices?

CHAPTER

23

Waste Management and Bioremediation

OBJECTIVE

After reading this chapter, you should be able to solve problems in the following areas:

- create compost mixes of known C:N from two types of starting material using a Pearson Square.
- determine biochemical oxygen demand (BOD).
- calculate the loading rates of gasoline wastes.

Overview

Waste management and bioremediation, the use of biological systems to decompose and remove toxic compounds from the soil environment, are two practical applications of a thorough understanding of soil science. If you understand something about the soil to which a waste is applied, and something about the capabilities of the microbes in that soil environment that actually perform the decomposition, then you are in a good position to understand the potential for using soil as a *waste removal engine* in an environmentally sound manner (Overcash & Pal, 1979).

Some of the basic concepts associated with waste management in soil are the ideal C:N ratio of the starting material, which we identified when we discussed mineralization in Chapter 17, and pretreatment of wastes to reduce pathogens, volume, and reactivity. Composting, a procedure for stimulating accelerated decomposition, is a time-honored method of accomplishing these goals and producing a final product that is suitable as a soil amendment. The Pearson Square is a simple method for mixing two wastes so that you end up with starting material for composting that has the appropriate C:N.

Reactivity of waste material added to soil or water generally refers, in a biological sense, to how much oxygen will be consumed during the waste's decomposition. If the material is too reactive (or labile), it means that oxygen consumption will also be rapid, which could lead to anaerobic conditions developing. Oxygen depletion is bad from two perspectives—first, anaerobic decomposition creates volatile compounds that smell really bad and cause an odor problem; second, in aquatic systems, if the oxygen depletion is too extreme, it can deprive fish and other aquatic animals of the oxygen that they require to live. You can anticipate the potential of a waste to cause oxygen depletion by measuring its biochemical (or biological) oxygen demand (BOD).

Finally, some of the most significant soil contaminants are gasoline products leaking from underground storage tanks (USTs). Four compounds—benzene, toluene, ethylbenzene, and xylene (BTEX)—are regulated because they are carcinogenic. Each compound will decompose in soil, but the half-life of each compound in soil is different, so it would be good to know just how much contaminated soil could be decomposed on a yearly basis to remove these toxins from the soil environment.

C:N Ratios and Composting

In composting, an ideal C:N ratio of the starting materials is between 15 and 30. If the ratio is much higher than 30, N immobilization occurs, which slows down the composting process, and if it is much lower than 15, excess N becomes available during mineralization, which can volatilize into the atmosphere and cause odor problems. Composting operations will mix various types of wastes together (so-called brown or green wastes, reflecting high and low C:N, respectively) to achieve the ideal ratio. How is that possible?

If we take the simplest situation, in which we mix only two materials together, you can use an aid called the **Pearson Square** to determine the appropriate ratios to use. The Pearson Square, which also has applications in generating feed rations, is a diagram in which the C:N of the starting materials is on the right axis, the desired C:N ratio is in the center of the diagram, and the appropriate ratios of the starting materials to use is on the left axis (Figure 23–1).

To use the square, rank the C:N of the two starting materials from highest to lowest and put them at the right-hand corners of the square. Put the desired C:N in the center of the square. Subtract the desired C:N from the highest C:N on the right-hand side (the upper right corner) and put the result in the bottom left-hand corner. Subtract the lowest C:N from the desired C:N and put the result in the upper left-hand corner. The numbers on the left-hand side represent the relative proportions of each starting material to use to obtain the desired C:N.

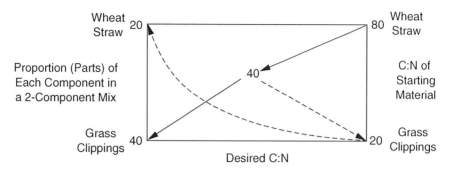

FIGURE 23–1 Application of the Pearson Square to determine optimal C:N ratios of two mixed compost materials.

Here's another way to look at it.

$$\text{High C:N (RT)} - \text{desired C:N (C)} = \longrightarrow \text{LB}$$

$$\text{Desired C:N (C)} - \text{low C:N (RB)} = \longrightarrow \text{LT}$$

$$\frac{\text{LT}}{\text{(LT + LB)}} = \text{fraction of total compost composed of low C:N material}$$

$$\frac{\text{LB}}{\text{(LT + LB)}} = \text{fraction of total compost composed of high C:N material}$$

where

RT = right top of the Pearson Square
RB = right bottom of the Pearson Square
LT = left top of the Pearson Square
LB = left bottom of the Pearson Square
C = center of the Pearson Square
$\longrightarrow$ = put the result in

Example 23–1

Prepare a compost mixture with a C:N of 40:1 if your starting materials are wheat straw (C:N = 80:1) and grass clippings (C:N = 20:1).

Solution

Step 1. Subtract 40 from 80 (Top right − desired C:N).
Step 2. Put result from Step 1 in the bottom left corner (=parts of grass clippings to use in the final mix).
Step 3. Subtract 20 from 40 (Desired C:N − lower right).
Step 4. Put result from Step 3 in the upper left corner (=parts wheat straw required in the final mix).
Step 5. The final compost mix should consist of 40 parts grass clippings and 20 parts wheat straw to get a final C:N ratio of 40:1.

What the Pearson Square lets you do is basically create the weighted average of the two starting materials.

Example 23–2

Show that the weighted average of 20 parts wheat straw and 40 parts grass clippings from Example 23–1 gives a C:N of 40:1.

Solution

$$\frac{\text{Parts straw}}{\text{Total parts}} \times \text{C:N straw} = \frac{20 \text{ parts straw}}{20 \text{ parts straw} + 40 \text{ parts clippings}} \times \frac{80}{1} = \frac{1600}{60}$$

$$\frac{\text{Parts clippings}}{\text{Total parts}} \times \text{C:N clippings}$$

$$= \frac{40 \text{ parts clippings}}{20 \text{ parts straw} + 40 \text{ parts clippings}} \times \frac{20}{1} = \frac{800}{60}$$

$$\frac{1600}{60} + \frac{800}{60} = \frac{2400}{60} = \frac{40}{1}$$

Biochemical Oxygen Demand

The **biochemical (or biological) oxygen demand** (BOD) is a standard technique for monitoring the oxygen demand of waste-contaminated waters, but it can also be applied to wastes added to soil for basically the same reasons—adding wastes that are too reactive (labile) will cause the oxygen in the soil atmosphere to be depleted by microbial activity. Oxygen depletion could cause anaerobic conditions to develop, which would create odorous volatile compounds. Oxygen depletion could also starve plant roots of the oxygen they need for growth.

In principle, measuring BOD is simple. One takes a known amount of sample (solid or liquid), adds it to a standardized reagent vessel full of buffer, measures the dissolved oxygen content by titration or with a BOD meter, incubates the vessel for 5 days at 20°C, and then measures the dissolved oxygen again. The difference in dissolved oxygen content between day 0 and day 5, accounting for dilution, is the BOD.

$$\text{BOD}_5 \ (\text{mg L}^{-1}) = \frac{D_1 - D_2}{P} \qquad \boxed{\textbf{23-1}}$$

where

D_1 = dissolved oxygen (DO) of the diluted sample on day 0
D_2 = DO of the diluted sample on day 5
P = decimal volumetric fraction of the sample used

In the case of a solid sample, you also have to take into account the extent to which the original solid sample was diluted. We went through some sample calculations of this type in Chapter 15 (Microbial Enumeration). The easiest approach for working with solids of varying density is to perform the initial dilution in a graduate cylinder so that an exact dilution factor can be determined based on the final volume and the dry weight of the material.

Example 23–3

Ten milliliters of a waste water sample is diluted to a final volume of 300 mL in a standard BOD bottle. On day 0, the dissolved oxygen content is 7.0 mg L^{-1}, and on day 5, the dissolved oxygen content is 3.0 mg L^{-1}. What is the BOD?

Solution

$$P = \frac{10 \text{ mL}}{300 \text{ mL}} = 0.033$$

$$BOD_5 \text{ (mg } L^{-1}) = \frac{D_1 - D_2}{P}$$

$$BOD_5 \text{ (mg } L^{-1}) = \frac{7.0 \text{ mg } L^{-1} - 3.0 \text{ mg } L^{-1}}{0.033} = 121.2 \text{ mg } L^{-1}$$

Example 23–4

Five grams of oven-dry equivalent material is diluted in buffer until the final volume is 100 mL. Thirty milliliters of the diluted sample is dispensed into a standard 300 mL BOD bottle and brought to volume. On day 0, the DO was 7.5 mg L^{-1}. On day 5, the DO was 1 mg L^{-1}. What is the BOD of this solid?

Solution

The dilution factor for the solid is 5 g/100 mL = 5 g/0.1 L. Any contributions of native moisture content and solid density to volume are taken into account in the final volume, which is given as 100 mL.

$$BOD_5 \text{ (mg } L^{-1}) = \left(\frac{D_1 - D_2}{P} \right) \div DF$$

$$BOD_5 \text{ (mg } g^{-1}) = \left(\frac{7.5 \text{ mg } L^{-1} - 1 \text{ mg } L^{-1}}{30/300} \right) \div 5 \text{ g/0.1 L} = 1.3 \text{ mg } g^{-1}$$

$$= 1300 \text{ mg kg}^{-1}$$

TABLE 23–1 Effect of Temperature and Pressure on Dissolved Oxygen Concentration (mg L^{-1})

Temperature		Pressure in mmHg(inHg)							
°F	°C	775(30.51)	760(29.93)	750(29.53)	725(28.45)	700(27.56)	675(26.57)	650(25.59)	625(24.61)
32	0	14.9	14.6	14.4	13.9	13.5	12.9	12.5	12.0
25.6	2	14.1	13.8	13.7	13.2	12.9	12.3	11.8	11.4
39.2	4	13.4	13.1	13.0	12.5	12.1	11.7	11.2	10.8
42.8	6	12.7	12.4	12.3	11.9	11.5	11.1	10.7	10.3
46.4	8	12.1	11.8	11.7	11.3	10.9	10.5	10.1	9.8
50.0	10	11.6	11.3	11.2	10.8	10.4	10.1	9.7	9.3
53.6	12	11.1	10.8	10.7	10.3	10.0	9.6	9.2	8.9
57.2	14	10.6	10.3	10.2	9.9	9.5	9.2	8.9	8.5
60.8	16	10.1	9.9	9.8	9.5	9.1	8.8	8.5	8.1
64.4	18	9.7	9.5	9.4	9.1	8.8	8.4	8.1	7.8
68.0	20	9.3	9.1	9.1	8.7	8.4	8.1	7.8	7.5
71.6	22	9.0	8.7	8.7	8.4	8.1	7.8	7.5	7.2
75.2	24	8.7	8.4	8.4	8.1	7.6	7.5	7.2	7.0
78.8	26	8.4	8.1	8.1	7.8	7.6	7.3	7.0	6.7
82.4	28	8.1	7.8	7.8	7.6	7.3	7.0	6.7	6.5
86.0	30	7.8	7.6	7.6	7.3	7.0	6.8	6.5	6.2
89.6	32	7.6	7.3	7.3	7.0	6.8	6.6	6.3	6.0
93.2	34	7.3	7.1	7.1	6.8	6.6	6.3	6.1	5.8
96.8	36	7.1	6.8	6.9	6.6	6.4	6.1	5.9	5.6
100.4	38	6.9	6.6	6.6	6.4	6.2	5.9	5.7	5.5
104.0	40	6.7	6.4	6.4	6.2	6.0	5.7	5.5	5.3

Note. HACH Water Analysis Handbook (3rd ed.). (2004). Loveland, CO: HACH.

A good rule of thumb is to keep the BOD added to soil at less than 500 lb acre^{-1} (560 kg/ha) per day.

The maximum DO on day 0 is determined by a combination of the temperature and pressure of the environment—two factors to which DO is sensitive, as Table 23–1 demonstrates. As temperature increases, and as pressure decreases, the DO content of water declines. This is one reason why it is important that the DO on day 0 and day 5 be measured under the same environmental conditions, and why most BOD meters are designed to compensate for changes in temperature and pressure.

It is important to remember that for the purposes of measuring BOD, the final DO content of the sample should not be <1 mg L^{-1} and that there should be at least a 2 mg L^{-1} difference between the starting and ending DO to ensure the accuracy of the measurement. This is regulated by the volume of sample used in the reagent vessel. For samples that have low BOD, a large sample volume is used; for samples with a high BOD a small sample is used. Table 23–2 illustrates this relationship. You can't measure BOD if the DO on day 5 is 0 mg L^{-1} because you have no idea whether the DO went to 0 on day 5 or sooner. If sooner, that would mean there is still some potential BOD in the sample that is not measured because all the oxygen has disappeared.

TABLE 23–2 Approximate Minimum Sample Volumes to Use for Waste Waters of Varying Strength (milliliters per 300 mL standard BOD bottle)

Sample Type	Approximate BOD (mg L^{-1})	mL of Sample to Use
Strong waste, raw and settled sewage	600	1
	300	2
	200	3
	150	4
	120	5
	100	6
	75	8
	60	10
Oxidized effluents	50	12
	40	15
	30	20
	20	30
	10	60
Polluted river water	6	100
	4	200
	2	300

Note. HACH Water Analysis Handbook (3rd ed.). (2004). Loveland, CO: HACH.

Bioremediation—Calculating Waste Loading Rates for Gasoline Spills

Let's say you have excavated soil around a leaking gas tank that is contaminated with 10,000 ppm (10,000 mg kg^{-1}) BTEX. What is the rate at which you can land apply this material? The maximum amount of waste you can add to the soil is determined using the formula (Environmental Protection Agency, 1983)

$$C_{yr} = \frac{1/2C_{crit}}{t_{1/2}}$$

23-2

where

C_{yr} = maximum application rate (lb acre^{-1} year^{-1} or kg ha^{-1} year^{-1})
C_{crit} = toxic concentration of the most toxic compound in the waste (lb acre^{-1} or kg ha^{-1})
$t_{1/2}$ = half-life of the longest-lived compound in the waste (usually expressed as days)

Half-lives of various compounds are given in Table 23–3.

Example 23–5

What is the maximum annual loading rate of soil contaminated with 10,000 ppm BTEX?

Solution

Step 1. First determine C_{crit}.

Assume toluene is the most toxic compound in the waste mix. 500 ppm = 0.0005 lb/lb (500 mg kg^{-1}), which is the level that has a partial sterilizing effect

TABLE 23–3 Half-Lives of Various Compounds

Compound	$t_{1/2}$ (days)
Benzene	1.8
Toluene	4.0
Ethylbenzene	6.4
p-xylene	10.8
m-xylene	7.5
o-xylene	14.7
Naphthalene	8.7
2-methylnaphthalene	9.9
1-methylnaphthalene	16.9

on microbes in soil. There are approximately 2×10^6 lb (2.24×10^6 kg ha^{-1}) of soil in the plow layer (excluding land that has a very high organic matter content).

$$(0.0005 \text{ lb/lb toluene}) \times (2 \times 10^6 \text{ lb acre}^{-1}) = 1000 \text{ lb toluene/acre}$$

$$(500 \text{ mg/kg toluene}) \times (2.24 \times 10^6 \text{ kg ha}^{-1}) = 1.12 \times 109 \text{ mg toluene/ha}$$

$$= 1120 \text{ kg toluene/ha}$$

Step 2. Determine the limiting half-life.

The half-life ($t_{1/2}$) is based on the most resistant fraction of the waste. If you look at Table 23–3, you will see that of the components of BTEX, *o*-xylene has the longest half-life—14.7 days. Put on a yearly basis this means

$$14.7 \text{ days} \div 365 \text{ days year}^{-1} = 0.0403 \text{ years}$$

Step 3. Determine maximum application rate.

We can now figure out the maximum amount of BTEX that can be added to soil.

$$C_{yr} = \frac{1/2 C_{crit}}{t_{1/2}}$$

$$C_{yr} = \frac{(1/2)(1000 \text{ lb acre}^{-1})}{0.043 \text{ years}} \qquad C_{yr} = \frac{(1/2)(1120 \text{ kg ha}^{-1})}{0.043 \text{ years}}$$

$$= 12,407 \text{ lb BTEX acre}^{-1} \text{ year}^{-1} \qquad = 13,023 \text{ kg ha}^{-1} \text{ year}^{-1}$$

Step 4. Determine the loading rate.

The loading rate is how much contaminated soil we can add to land. To figure this out we use the formula

$$\text{Mass soil applied} = \frac{C_{yr}}{C_{waste}} \qquad \textbf{23-3}$$

The concentration of BTEX in the soil is 10,000 ppm or 0.01 lb BTEX per pound soil (10,000 mg BTEX per kilogram soil). We just calculated the maximum amount we could apply to soil (C_{yr}), which was equal to 12,407 lb BTEX acre^{-1} year^1 (13,023 kg ha^{-1} year^{-1}). So, the amount of soil contaminated at this level that we could land apply is

$$\text{Mass soil applied} = \frac{C_{yr}}{C_{waste}}$$

$$\text{lb soil applied} = \frac{12,407 \text{ lb BTEX acre}^{-1} \text{ year}^{-1}}{0.01 \text{ lb BTEX/lb soil}} = 1,240,700 \text{ lb soil/acre/year}$$

$$\text{kg soil applied} = \frac{13,023 \text{ kg BTEX ha}^{-1} \text{ year}^{-1}}{10,000 \text{ mg BTEX/kg soil}} = 1,302,300 \text{ kg soil/ha/year}$$

References

Environmental Protection Agency. (1983). *Hazardous waste land treatment.* (Rev. ed., EPA SW-874). Washington, DC: Author.

HACH Water Analysis Handbook (3rd ed.). (2004). Loveland, CO: HACH.

Overcash, M. R., & Pal, D. (1979). *Design of land treatment systems for industrial wastes—theory and practice.* Ann Arbor, MI: Ann Arbor Science.

Sample Problems

Using the Pearson Square

For Questions 1 to 10, use the information for these wastes in your calculations.

Cattle manure:	1.5% N, 0.5% P, 0.6% K, 1.0% Ca, 0.3% Mg, C:N = 18
Alfalfa:	2.4% N, 0.2% P, 1.8% K, 1.4% Ca, 3.9% Mg, C:N = 15
Corn stover:	0.9% N, 0.1% P, 1.2% K, 0.4% Ca, 0.1% Mg, C:N = 42
Oat straw:	0.6% N, 0.1% P, 1.2% K, 0.6% Ca, 0.1% Mg, C:N = 90
Rice hulls:	0.6% N, 0.1% P, 0.7% K, 0.4% Ca, 0.1% Mg, C:N = 85
Pine sawdust:	0.1% N, <0.1% P, 0.1% K, 0.1% Ca, <0.1% Mg, C:N = 225
Mixed green weeds:	2.3% N, 0.3% P, 1.3% K, 0.1% Ca, <0.1% Mg, C:N = 21
Chicken Manure:	4.5% N, 0.8% P, 0.7% K, 1.8% Ca, 0.4% Mg, C:N = 7
Fish Processing Waste:	9.0% N, 7.0% P, 0.8% K, 1.4% Ca, 0.1% Mg, C:N = 4
Lime-Stabilized Biosolids:	3.6% N, 1.2% P, 0.4% K, 3.6% Ca, 0.4% Mg, C:N = 14
Starch Processing Waste:	0.1% N, <0.1% P, <0.1% K, <0.1% Ca, <0.1% Mg, C:N = 312
Mixed Papermill Sludge:	0.9% N, 0.1% P, <0.1% K, 6.9% Ca, 0.3% Mg, C:N = 61

Using the Pearson Square as an aid, prepare a compost that has a C:N ratio of 35:1 for each combination of wastes.

1. chicken manure + corn stover
2. fish processing waste + pine sawdust
3. lime-stabilized biosolids + oat straw

4. starch processing waste + mixed green weeds
5. mixed papermill sludge + mixed green weeds
6. chicken manure + rice hulls
7. fish processing waste + corn stover
8. lime-stabilized biosolids + pine sawdust
9. starch processing wastes + alfalfa
10. mixed papermill sludge + cattle manure

Calculating BOD

1. What is the BOD of 15 mL of a liquid sample if the starting BOD was 9.3 and the final BOD was 5.0, and a standard 300 mL BOD bottle was used?
2. What is the BOD of 100 mL of polluted river water added to a standard 300 mL BOD bottle if the starting BOD was 8.2 and the final BOD was 6.6?
3. If the final measured BOD of runoff from a beef cattle feedlot was 2201 mg L^{-1}, and the starting DO was 8.7 mg L^{-1}, how much would you have to dilute the original sample to ensure that the DO doesn't decline by more than 50 percent during the incubation period?
4. What is the BOD of 2 g of oven-dry material if it is diluted to a total volume of 1000 mL, and 5 mL of this dilution causes a total DO change of 3.5 mg L^{-1} after incubation?
5. If the BOD of fresh cattle slurry from a lagoon is 16.1 g L^{-1}, how many liters can you apply to soil without exceeding a loading rate of 400 kg BOD ha^{-1}?
6. Assume you can apply a maximum amount of BOD of 500 kg ha^{-1} per day. If a liquid waste has a measured BOD of 2000 mg O_2 L^{-1}, what volume of waste can you apply to soil?
7. If a water sample from feedlot runoff was diluted 1000 times and used to measure BOD, what was the BOD if at day 0 the dissolved oxygen was 8 mg L^{-1} and at day 5 it was 0 mg L^{-1}?
8. Prior to land application, 1 g of vegetable waste was diluted 500-fold in water and the resulting BOD was determined. The initial DO was 7.0 mg L^{-1} and the final DO was 2 mg L^{-1}. How much O_2 would be required to meet the biochemical demand for decomposing this waste in soil if 50 Mg (50,000 kg) were land applied?

Bioremediation

1. Assume you have a gasoline-contaminated soil containing 6500 ppm BTEX. The C_{crit} for toluene will be assumed to be 500 ppm. Assume that o-xylene is the most resistant fraction of your waste. What is the loading rate of contaminated soil (in kg ha^{-1}) that you can apply on a yearly basis?
2. Assume you have a gasoline-contaminated soil with 10,000 ppm BTEX. The C_{crit} for toluene, the most toxic constituent will be assumed to be 500 ppm. Assume that for this sample p-xylene is the most resistant constituent. What is the loading rate of contaminated soil that you can apply on a yearly basis?

3. You have convinced your state's Drug Enforcement Agency (DEA) to let you produce industrial hemp as long as it is grown on land amended with toluene wastes. If the C_{crit} for toluene is 5 kg ha^{-1} when industrial hemp is grown (i.e., if the amount of toluene exceeds 5 kg ha^{-1}, your crop dies) and you are receiving 1 Mg of toluene to dispose of on a yearly basis, how much land do you need to dispose of the toluene and still grow your industrial hemp?

4. The C_{crit} for ethanol is 46,000 ppm and it has a half-life of 1 day. How much 180 proof alcohol from your home still can you dispose of in your backyard (1 ha in size) in a year without worrying about revenue agents discovering you?

5. Assume you have a soil contaminated with 77,500 ppm BTEX. The C_{crit} for the most toxic component of this waste is assumed to be 400 ppm. Assuming that o-xylene is the most recent constituent, with a half-life of 14.5 days, what would be the loading rate of contaminated soil (in kg ha^{-1}) that you could apply on a yearly basis assuming that no more than 20 percent of the BTEX volatilizes?

Section VI
Data Analysis and Sampling

24

Descriptive Statistics

OBJECTIVE

The objective of this chapter is to introduce descriptive statistics as a tool to summarize soil science data. After completing this chapter, you will be able to

- characterize data sets.
- construct frequency tables.
- calculate various measures of central tendency and dispersion.

Overview

Data collected in soil science should be summarized and interpreted before its presentation. While most people can draw their own conclusions from raw data, it is clearly helpful for them to have a summary of the data prepared by established techniques, so that they don't have to use their time summarizing the data for themselves.

Descriptive statistics refers to summarizing the data logically and portraying it as simple statistical means (averages), or in charts or tables. Inferential statistics refers to making a statement about whether the data are a probable occurrence based on some preset hypothesis. In this chapter, you will look at types of data and some basic descriptive statistics. In Chapter 25, you will use error analysis to tell whether some data values are believable. In Chapter 26, you will look at inferential statistics and how to describe the relationship between sets of data that may be related. In Chapter 27, you will get some experience in how statistics can assist soil sampling.

Types of Data

Data can be quantitative (numerical), in which case it is described as continuous or discrete. Continuous observations, for example, are observations for which numerical values fall within a range, such as the weight of different soil particles. Discrete observations, on the other hand, are observations in which the measured value changes in distinct steps, usually in whole numbers, for example, the number of persons in a family (hence, the long-standing joke that the average family consists of 2.58 children; this is a misapplication of averaging because persons represent a discrete value).

Data can also be qualitative or categorical. These are observations that have no numeric value. Qualitative data can be divided into two classes—ordinal and nominal. Ordinal data are data that can be ordered or categorized by some criterion such as hue (for a color) or intensity of symptom (none, moderate, or severe) for ranking disease. Nominal data are observations that have no inherent rank. Color itself, for example, would be a nominal observation.

Example 24–1

For the following data sets, identify what type of data (continuous, discrete, ordinal, or nominal) they represent.

A. 8 eggs, 4 eggs, 7 eggs, 10 eggs
B. birch, larch, oak, maple
C. transparent, opaque, translucent, dark
D. 3.6, 8.4, 3.9, 4.5

Solution

A. Discrete, each group of eggs has a distinct number.
B. Ordinal, each tree is a distinct type.
C. Nominal, the terms refer to the intensity of transmitted light.
D. Continuous, each number can be considered a potential value within a range.

Descriptive Statistics

Frequency Tables

A sample frequency table provides a quick and simple summary of data based on how often particular data appear. Presumably, the data values are common if they occur frequently and uncommon if they occur infrequently. This presentation method lends itself well to nominal data, but it can also be applied to other types of data. The historical importance of frequency tables can't be overestimated. Remember that Gregor Mendel, the Austrian monk considered to be the father of modern genetics, developed his ideas about genes based on the frequency with which nominal traits (such as smooth or wrinkled seeds) occurred in different generations of pea plants.

Example 24–2

In a test plot, students planted 144 seeds from hybrid corn with leaves of medium width and obtained the following results: 36 broad-leafed plants, 76 medium-leafed plants, 32 narrow leafed plants. Summarize the data in a frequency table.

Solution

Class (Leaf width)	Frequency	Relative frequency (%)
Narrow	32	22
Medium	76	53
Broad	36	25
Total	144	100

The **relative frequency** is the frequency of an individual observation relative to the total number of observations. Relative frequency is useful when comparing observation sets within similar classes, but which contain different numbers of observations.

   If you have ever been graded on a curve, then your grade (a discrete value) was likely arranged in a frequency table.

Example 24–3

For an introductory soil science class, the following distribution of grades (%) were obtained after the first exam: 79, 88, 83, 96, 79, 75, 92, 96, 83, 83, 88, 88, 88, 79, 88, 88, 88, 92, 100, 88, 92, 83, 96, 88, 92, 88, 71, 63, 83, 100, 96. Summarize the data using a frequency table.

Solution

Grade (%)	Frequency	Relative frequency (%)
63	1	3
71	1	3
75	1	3
79	3	9
83	6	18
88	10	30
92	5	15
96	4	12
100	2	6
Total	33	99
		(Due to rounding)

Obviously, if the most frequent grade was going to be assigned a C, then 88 percent might only get you a C after this exam, since it occurs most frequently.

Continuous data don't lend themselves well to frequency tables because they often contain too many individual observations with insufficient frequency to be of much use. To get around this, measurements are typically grouped into classes. The classes should be numerous enough to show trends, but not so numerous that it reduces the frequency within each class to a meaningless value. Each data point should fall in only one class. A good formula to use in determining the optimum number of classes for a continuous data set is calculated by Equation 24-1.

$$2^{k-1} \leq n \leq 2^k \qquad \textbf{24-1}$$

where

k = number of classes
n = total number of observations

Example 24–4

For the following soil temperature measurements (°C), determine the appropriate number of classes to use, and develop a suitable frequency table: 16.9, 16.0, 15.4, 17.6, 18.9, 18.0, 16.0, 15.6, 17.9, 17.3, 17.0, 17.0, 17.0, 17.5, 17.4, 17.4, 17.6, 16.9, 16.5, 17.5, 17.5, 17.2, 17.2, 15.0, 17.6, 17.8, 16.5, 16.9, 14.7, 16.9, 16.6, 18.9, 17.0, 16.0, 15.0, 16.0, 17.0, 16.0, 17.5, 13.0, 16.0, 17.8, 16.7, 16.8, 17.8, 15.4, 17.6, 17.2, 15.2, 17.2.

Solution

The number of observations (n) = 50.
Use k as an exponent to bracket n. Play with the value of 2^k until $2^k > n$, therefore $2^{k-1} \leq n \leq 2^k$.

$$2^{6-1} \leq 50 \leq 2^6$$

The appropriate number of classes to use is 6. The range of the data is from 13.0 to 18.9°C, a difference of 5.9.
5.9/6 = 0.98, approximately 1.0, so each class should span about 1°C.

Class boundary	Frequency	Relative frequency (%)
13.0–13.9	1	2
14.0–14.9	1	2
15.0–15.9	6	12
16.0–16.9	15	30
17.0–17.9	24	48
18.0–18.9	3	6
Total	50	100

Measures of Central Tendency

For continuous data, frequency of appearance is less important than a definitive measure of some average or central value that could be used to summarize the whole data set. There are several ways of measuring central tendency. The arithmetic mean ($\overline{X}$) is the sum of all observations in a data set (ΣX) divided by the number of observations (n) (remember that Σ is shorthand for writing "the sum of").

$$\overline{X} = \frac{\sum X}{n}$$

24-2

Example 24–5

What is the arithmetic mean of the following data series: 2, 8, 4, 6, 10, 2, 10, 5?

Solution

$$\overline{X} = \frac{\sum 2 + 8 + 4 + 6 + 10 + 2 + 10 + 5}{8}$$

$$\overline{X} = \frac{47}{8} = 5.9 \text{ (expressed as two significant figures)}$$

The geometric mean (G) is frequently calculated during water quality monitoring. It is also used to calculate relative values for index numbers and in averaging ratios and rates such as investment returns or population growth (Steele & Torrie, 1980). Equation 24-3 describes the calculation for a geometric mean.

$$G = \sqrt[n]{X_1 X_2 \ldots X_n}$$
$$= (X_1 X_2 \ldots X_n)^{1/n}$$

24-3

Example 24–6

What is the geometric mean (G) of the following data series: 2, 8, 4, 6, 10, 2, 10, 5?

Solution

$$G = \sqrt[8]{(2)(8)(4)(6)(10)(2)(10)(5)}$$

$$G = (384000)^{1/8}$$

$$= 5.0$$

Note that the geometric mean is only applicable to data sets consisting of positive numbers (because you can't take the root of a negative number).

The weighted average mean is used when there are several means being averaged and each is based on different numbers of observations. Let's say you wanted to find the weighted average of heights that had been grouped by class in a frequency table. Weighted averages are also used, for example, when nutrient concentrations are averaged over time in surface and ground water samples and each nutrient concentration is based on a different water volume. The equation for a weighted mean is

$$\overline{X}_W = \frac{\sum W_n X_n}{\sum W_n}$$

24-4

where W represents the number of observations or volume, etc.

Example 24–7

What is the weighted average nitrate (NO_3^-) concentration in runoff over a 15-min interval given the following data?

Time (min)	Flow (mL)	NO_3^- Concentration (ppm)
0	250	4.25
5	800	4.60
10	900	4.70
15	1000	4.10

Solution

$$\overline{X}_W = \frac{\sum W_n X_n}{\sum W_n}$$

Time (min)	Flow (mL)	NO_3^- Concentration (ppm)	$\sum W_n X_n$
0	250	4.25	1062.5
5	800	4.60	3680.0
10	900	4.70	4230.0
15	1000	4.10	4100.0
	$\sum W_n = 2950$		$\sum W_n X_n = 13,072.5$

$$\overline{X}_W = 13,072.5/2950$$

$$= 4.43 \text{ ppm}$$

Moving or rolling averages are frequently utilized in applications where continuous measurements are made, but some time elapses before the data values are actually obtained. A good example is biochemical oxygen demand (BOD) measurements, where five to seven days elapse between the time samples are collected and results are obtained. Moving averages are also useful in continual monitoring of water samples that are subject to periodic fluctuations caused by external events, such as storms. A moving average is calculated as follows:

$$\overline{X}_1 = \sum \left(\frac{X_1 + X_2 \ldots + X_n}{n} \right)$$

$$\overline{X}_2 = \sum \left(\frac{X_2 + X_3 \ldots + X_{n+1}}{n} \right)$$

$$\overline{X}_3 = \sum \left(\frac{X_3 + X_4 \ldots + X_{n+2}}{n} \right)$$

and so on.

Example 24–8

Calculate the 5-day moving average for the following BOD data set:

Day	BOD (mg L^{-1})
1	50
2	110
3	70
4	70
5	60
6	80
7	50
8	50
9	50
10	70

Solution

You can create your 5-day moving average after the fifth daily measurement.

$$50 + 110 + 70 + 70 + 60 = 360/5 = 72$$

Subsequent moving averages are calculated by adding the data from the most recent measurement available and dropping the data from the least recent measurement. So, for subsequent days, the moving average is

Day 6 $110 + 70 + 70 + 60 + 80 = 390/5 = 78$
Day 7 $70 + 70 + 60 + 80 + 50 = 330/5 = 66$
Day 8 $70 + 60 + 80 + 50 + 50 = 310/5 = 62$
Day 9 $60 + 80 + 50 + 50 + 50 = 290/5 = 58$
Day 10 $80 + 50 + 50 + 50 + 70 = 300/5 = 60$

The **median** is the middle value of a data series that has been ordered by rank. If there are an odd number of values in the data series, the median is the middle value. If there is an even number of values, the median value is the average of the two middle data points.

Example 24–9

What is the median of the following data series: 2, 8, 4, 6, 10, 2, 10, 5?

Solution

First, order the data series by rank—2, 2, 4, 5, 6, 8, 10, 10.

Because there are an even number of data points, the median is the average of the two middle data points.

$$(5 + 6)/2 = 5.5 \text{ (6 expressed as one significant figure)}$$

Because one or two extremely high or low values can have undue effect on the arithmetic mean, the median value is often used when the data can potentially be skewed. For example, housing costs and national exam scores are frequently expressed as median values because expensive homes or very bright students inflate the average values.

The **mode** is the most frequently occurring value in a data set. There is no mode if no data value occurs with any greater frequency than other members of the data set. Some data sets are called *bimodal* because they have two sets of data values that cluster together.

Measures of Dispersion

In many instances, you want to know the central tendency of a data series, and how much those data are dispersed, which indicates something about the variability of the data (its precision).

The **range** is the difference between the highest and lowest values of a data series. The range is not very useful if an extremely high or low value is present while the other values are relatively clustered.

Example 24–10

What is the range of the following data series: 2, 8, 4, 6, 10, 2, 10, 5?

Solution

The highest value is 10, the lowest value is 2. $10 - 2 = 8$. So, the range is 8.

A more useful measure of dispersion is the **variance,** a mathematical description of how dispersed data values are about a central value. Small variances represent values that are clustered around a central value while large variances represent values that are dispersed. The equation for calculating variances is

$$s^2 = \frac{\sum (X - \overline{X})^2}{df}$$

24-5

where

s^2 = variance
X = individual data value
$\overline{X}$ = arithmetic mean
df = degrees of freedom (the number of observations − 1)

As a practical means of calculating variance, however, the following equation is used:

$$s^2 = \frac{\sum X^2 - \frac{(\sum X)^2}{n}}{n - 1}$$ **24-6**

In practice, most calculators have statistical programs embedded in them that will allow you to calculate variance directly.

Example 24–11

Calculate the variance of the following data set: 2, 8, 4, 6, 10, 2, 10, 5.

Solution

X	X^2
2	4
8	64
4	16
6	36
10	100
2	4
10	100
5	25
$\sum X = 47$	$\sum X^2 = 349$

$$s^2 = \frac{349 - \frac{(47)^2}{8}}{8 - 1}$$

$$s^2 = 10.4$$

The **sample variance** (s^2) is in units squared, for example, $(\text{mg L}^{-1})^2$, which is not a particularly useful way of expressing it. Consequently, the most common method of expressing variance is to use the **standard deviation** (s), which is simply the square root of the variance.

$$\text{Standard deviation} = s = \sqrt{s^2}$$ **24-7**

Example 24–12

What is the standard deviation of the following data set: 2, 8, 4, 6, 10, 2, 10, 5?

Solution

Since you have already calculated the variance (see Example 24–11) ($s^2 = 10.4$), the standard deviation is $\sqrt{10.4} = 3.2$.

When replicate plots are used, and several measurements are made from each plot to identify the arithmetic mean value of each plot, then it is appropriate to report the dispersion of the data as the **standard deviation of the mean** or **standard error** (sem).

$$\text{Standard error of the mean (sem)} = \frac{s}{\sqrt{n}} \qquad \boxed{24\text{-}8}$$

Example 24–13

Assuming the data series 2, 8, 4, 6, 10, 2, 10, 5 represents the mean values of eight replicate measurements for different plots, calculate the standard error of the mean.

Solution

Since you have already calculated the standard deviation for this data set ($s = 3.2$), the standard error of the mean is

$$\frac{s}{\sqrt{8}} = 1.13$$

The magnitude of standard deviations depends on the magnitude of values used to create them. This means that measurements having a large magnitude will appear to have much greater dispersion than measurements with small magnitudes, even though the actual dispersion of data around the central mean in each data set is approximately the same.

To allow a comparison of standard deviations created from data sets of different magnitude, the relative standard deviation or **coefficient of variation** (CV) is used. The CV (as it is most commonly known) is simply the ratio of the sample standard deviation to the arithmetic mean expressed as a percentage. Low CVs indicate that the data are clustered; high CVs indicate that the data are dispersed.

$$CV = \left(\frac{s}{\overline{X}}\right)(100) \qquad \boxed{24\text{-}9}$$

Example 24–14

What is the standard deviation and CV of the following two data sets A and B?

A. 2, 8, 4, 6, 10, 2, 10, 5
B. 20, 80, 40, 60, 100, 20, 100, 50

Solution

The mean of sample A is 5.9. The mean of sample B is 58.8. You have already calculated the standard deviation for data set A (3.2). Data set B, which differs in each case by a factor of 10 (or one order of magnitude), has a standard deviation of 32.3. However, the CV of each data set

A. (3.2/5.9)100 = 54%
B. (32.3/58.8)100 = 55%

is virtually identical, meaning both have the same precision of measurement.

Reference

Steele, R. G., & Torrie, J. H. (1980). *Principles and procedures of statistics* (2nd ed.). New York: McGraw-Hill.

Sample Problems

1. Are the following types of data quantitative or categorical; discrete, continuous, ordinal, or nominal?

    A. the rate at which bacteria respond to external stimuli
    B. the number of legs on a spider
    C. the yield from 72 farms in Iowa
    D. the color of various grass species
    E. the registration preferences of a group of voters

2. Complete a frequency table for the following data listing soil series that were found in various farms in Kentucky: Al, Pt, Ed, No, Fr, Ot, Ed, No, Bo, Fr, Fw, Bo, No, Fr, Ed, No, Pt, La, No, Ed, Nr, Ot, Ot, Bo, Ed, Nr, Mc, Fr, Fr, Fr, La, Mc, Mc, Ot, No, Nf, Nr, Fr, La, Fw, La, Mc, La, Mc, Ma, Bo, Mc, Ot, Al.
    Soils key: Allegheny silt loam (Al), Boonesboro silt loam (Bo), Dunning silty clay loam (Du), Eden silty clay loam (Ed), Fairmount-Rock outcrop (Fr), Faywood silt loam (Fw), Lawrence silt loam (La), Lindside silt loam (Ln), Lowell silt loam (Lo), Maury silt loam (Ma), McAfee silt loam (Mc), Newark silt loam (Nr), Nicholson silt loam (Nf), Nolin silt loam (No), Otwell silt loam (Ot), Pitts-Dumps complex (Pt).

3. Develop a frequency table for the following data set of oven-dry soil weights (g/sample): 165, 183, 168, 166, 155, 176, 171, 181, 176, 178, 173, 153, 152, 165, 174, 175, 173, 157, 170, 180, 178, 190, 169, 170, 165, 180, 168, 185, 171, 187.

4. What is the arithmetic mean of the data set in Question 3?

5. What is the geometric mean of the following data series: 5, 8, 6, 7, 2, 4, 12, 10?

6. What is the weighted average fecal coliform concentration in the following data?

Fecal coliform conc. (CFU/100 mL)	Sample size (mL)
5	100
10	50
20	75
30	200
4	75

7. What is the median value and mode for the data set in Question 3?
8. What is the variance and standard deviation of the data in Question 3?
9. The following data were collected from an experiment examining CO_2 respiration from soil:

Treatment	Plot no.	CO_2 evolution (mg CO_2 m^{-2})				
+N	1	50	45	40	27	46
	2	49	48	45	45	46
	3	61	50	52	45	43
−N	4	75	74	85	73	72
	5	73	73	79	77	76
	6	86	80	75	76	81

Calculate the standard error for these data.

10. What is the CV for the following data series: 15, 12, 42, 26, 19, 21, 18?

25

Error Analysis

OBJECTIVE

In this chapter you will learn how to

- determine if data are normally distributed.
- detect outliers.

Overview

Everyone makes mistakes. Sometimes sample analysis errors generate a data value suspiciously inconsistent with other values collected at the same time. These data values are known as outliers. The purpose of this chapter is to introduce several mathematical approaches that you can use to eliminate suspicious data values with a clear conscience.

Normally Distributed Data

When soil scientists say data are normally distributed, they are implying that most values will fall around a central mean, and that progressively fewer samples will be found that are much greater or much less than the central mean.

Normal distribution sometimes does not occur, as when one or more data values cause the data to be skewed relative to the presumptive central mean. Scientists frequently resort to taking the logarithmic value of each data point (called a log transformation). These values will typically fit a normal curve and a statistical analysis can proceed.

In normally distributed data (Figure 25–1) about 68 percent of all data values will be within one standard deviation of the mean, and 95 percent will be

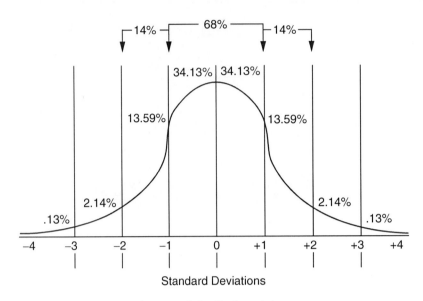

FIGURE 25–1 Plot of a curve for normal distribution of data.

within two standard deviations of the mean. Virtually all measurements should be within three standard deviations of the mean.

Detecting Outliers with the Z-Score

If you have a large number of samples (30 or more), you can use what is called a **Z-score** to detect outliers. Large samples are required because you want to approximate a normal distribution of the measured property. The formula for calculating the Z-score is

$$Z = \frac{(X - \overline{X})}{s}$$

25-1

where

Z = Z-score
X = individual data value
$\overline{X}$ = arithmetic mean
s = sample standard deviation

If the Z-score is <-4 or >4, the sample should be presumed to be an outlier because it is at least four standard deviations greater or less than the arithmetic mean, and in normally distributed data, virtually all sample values occur within ± 3 standard deviations of the mean.

Example 25–1

For the following data set, determine whether there are any outliers: 12, 5, 20, 22, 17, 15, 19, 24, 16, 15, 16, 18, 13, 14, 17, 14, 17, 22, 17, 19, 27, 19, 19, 18, 16, 13, 18, 19, 18, 20.

Solution

The arithmetic mean $(\overline{X}) = \Sigma X/n = 519/30 = 17$, the standard deviation $(s) = 4$, $n = 30$. So, if the difference between a sample and the arithmetic average is 16 or more, this sample is likely to be an outlier because it would give a Z-score of ± 4. The range of data is from 5 to 27.

$$\frac{(5 - 17)}{4} = -3$$

$$\frac{(27 - 17)}{4} = 2.5$$

Neither extreme is an outlier based on the predesignated cutoff point for the Z-score, so none of the values within the extreme are outliers either, and all the values are at least plausible based on the criterion for outliers we used.

Tests of Normality

Of course, you assumed a normal distribution of the data in Example 25–1. If the data were skewed, you might suspect the validity of the result. A quick measure of the skewness of the data set is

$$\frac{3\ (\text{Sample mean} - \text{sample median})}{\text{Sample standard deviation}}$$

 25-2

For perfectly symmetrical data (a perfect normal curve), the measure of skewness is zero because the mean and median are identical.

Example 25–2

For the data set in Example 25–1, determine its skewness.

Solution

If the data are ranked, the median value is 17.5. Since the mean value is 17.0, the measure of skewness is

$$\frac{3\ (17.0 - 17.5)}{4} = -0.4$$

This is only slightly skewed data. So, the assumption that you could use statistics intended for normally distributed data was probably valid.

The Box Plot Test

Another test for outliers is the **box plot test.** In this test, you actually determine five distinct statistical values of a data set—smallest value, largest value, lower quartile, upper quartile, and median. The smallest and largest values mark the range of the data set and are given. You already know how to determine the median value of a data set (Chapter 24). The lower, or first, quartile (Q_1) is calculated as

$$Q_1 = \frac{1\,(n + 1)}{4} \qquad \text{25-3}$$

For the data set in Example 25–1, $n = 30$, so the first quartile (the first fourth of the data) is the 7.75th data point in the ranked data set. In this case, 0.75 of the way between two measurements that are both 15, the first quartile has a value of 15. The third quartile (Q_3) is similarly calculated as

$$\frac{Q_3 = 3\,(n + 1)}{4} \qquad \text{25-4}$$

If $n = 30$, $Q_3 = 3\,(30 + 1)/4 = 23.25$. So, the third quartile ends 0.25th of the way between the 23rd and 24th data values. For the data in Example 25–1, this would be a value of 19.25.

The interquartile range (IQR) is the difference between the first and the third quartiles.

$$\text{IQR} = Q_3 - Q_1$$

$$= 19.25 - 15$$

$$= 4.25 \qquad \text{25-5}$$

In constructing a box plot, the box (IQR) contains 50 percent of all the data values. An inner fence is defined as

$$Q_1 \pm 1.5\,(\text{IQR}) \qquad \text{25-6}$$

$$15 \pm 1.5\,(4.25) = 8.6 \text{ or } 21.4$$

An outer fence is defined as

$$Q_3 \pm 3.0\,(\text{IQR}) \qquad \text{25-7}$$

$$15 \pm 3.0\,(4.25) = 6.5 \text{ or } 32.0$$

Putting the major parameters in place creates the following plot (Figure 25–2). Values that fall between the inner and outer fences are suspected outliers, while values that fall outside the outer fences are clearly outliers.

Example 25–3

Are there any data values in the data set from Example 25–1 that are outliers or suspected outliers?

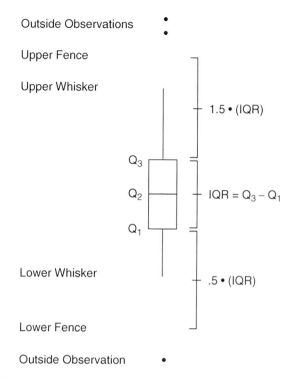

FIGURE 25–2 Diagram of a box plot.

Solution

Based on the calculations for the box plot test, values <6.5 or >32 are clearly outliers. The data value of 5 is an outlier by this criterion. Data that fall between 6.5 and 8.6 or 21.4 and 32.0 are suspected outliers. Based on these criteria, all values >22 in this data set are suspected outliers.

Sample Problems

Answer the following questions based on the following data set, which records the number of earthworms in a no-tillage field: 165, 183, 168, 166, 155, 176, 171, 181, 176, 178, 169, 161, 177, 166, 160, 184, 172, 169, 164, 160, 174, 154, 152, 165, 174, 175, 173, 158, 170, 180, 172, 175, 180, 146, 169, 164, 175, 194, 185, 164, 178, 190, 169, 170, 165, 180, 168, 185, 171, 187.

1. What are the mean, median, and standard deviation of this data set?
2. Is the data normally distributed?
3. Use the Z-score to detect any outliers.
4. Construct a box plot to detect any outliers.

CHAPTER

26

Hypothesis Testing and Inferential Statistics

OBJECTIVE

In this chapter you will learn how to calculate the most basic statistics used in comparing, analyzing, and drawing inferences from data sets with the help of the following:

- hypothesis testing with the t test
- linear regression
- correlation
- confidence intervals
- variance testing

Overview

Hypotheses are educated guesses about whether an observed result will be comparable to an expected result (Masterson & Slowinski, 1970). If you knew everything about a particular relationship, predicting the expected result would be relatively easy. Unfortunately, soil scientists never have complete knowledge of their subjects. They must infer the anticipated result based on sampling a small subset. Consequently, inferential statistics is used, along with hypothesis testing, to predict or generalize from a specific result—obtained by observation or experimentation—to overall phenomenon (Atlas & Bartha, 1993).

Hypothesis Testing and Levels of Significance

One of the most common hypothesis tests is whether a population (or other measurable property) is affected by two different treatments (Rees, 1989). For example, do plants grow faster if C is added to their environment compared to plants growing without additional C.

A null or starting hypothesis (H_0) would be that there was no difference between the two treatments (let's call them Trt_A and Trt_B, respectively). The null hypothesis would be that $Trt_A = Trt_B$. An alternative hypothesis (H_1) is that there is a difference between the two treatments ($Trt_A \neq Trt_B$). If you remember from Chapters 24 and 25, however, scientists estimate true means based on multiple samples, and there is some variability between measurements that cluster around a central mean. So, the mean response of Trt_A may not be equal to Trt_B, but the difference may be so small that it really only reflects the variability of the measurements used to estimate the true values of Trt_A and Trt_B.

To accept or reject the null hypothesis means taking a chance that your conclusions are correct based on the variability of the data. The level of significance (the alpha or α value) is the probability that you will incorrectly guess that there is a difference between Trt_A and Trt_B (reject H_0) when, in fact, there isn't. The level of significance is usually reported as $\alpha = 0.05$, or 0.01, which means that you will guess wrong either 5 or 1 percent of the time.

In football, quarterbacks know that during a pass play two out of three things can go wrong (the pass can be caught, but it can also be incomplete or intercepted). Likewise, statisticians know that in hypothesis testing, three out of four decisions can be incorrect or erroneous, as Table 26–1 illustrates (Carmer & Walker, 1988).

A Type I error, which is the α value, is the probability that you will decide there is a treatment difference when there really is not. Scientists testing for significant differences try to keep the α value as low as possible to reduce the

TABLE 26–1 Outcomes of Testing Hypotheses about Treatment Differences in Terms of Plant Growth with (Trt_A) and without (Trt_B) Carbon

Decision Based on Observed Differences	True Result		
	$Trt_A < Trt_B$	$Trt_A = Trt_B$	$Trt_A > Trt_B$
1. $Trt_A < Trt_B$	Correct decision	Type I error	Type III error
2. $Trt_A = Trt_B$	Type II error	Correct decision	Type II error
3. $Trt_A > Trt_B$	Type III error	Type I error	Correct decision

Note. Carmer, S. G., & Walker, W. M. (1988). Significance from a statistician's viewpoint. *Journal of Production Agriculture, 1,* 27–33.

significance of such an error. It is a conservative approach, often for good reason. Here is an example. Perhaps your experiments suggested that there was a significant difference in sewage treatment by Method A compared to Method B, when in fact there was not. If Method A was twice as expensive to implement as Method B, then whoever used your conclusions as the basis for their economic decisions could be paying a lot of extra money for no extra benefit.

A Type II error occurs when the null hypothesis is accepted when it really is false. If you have a very small α value, then the potential for a Type II error increases. In the example of the sewage treatment methods, whoever used your conclusions might miss out on a potentially better treatment process.

A Type III error is called a *reverse decision,* and is the probability that you will mistakenly attribute a real treatment difference to the wrong treatment. Type III errors never have an α value $> \alpha/2$ (Carmer & Walker, 1988).

Inferential Statistics—the *t*-Test

Once you've decided on the level of significance you're willing to accept for your experiment or observations (and this should always have been decided before you start), then you are ready to perform a statistical analysis of the data (Kiazolu, Mrema, & Sebolai, 1994). Generally, this is apt to be the comparison of two samples that have been treated differently.

For small samples, in which the true population variance is unknown, a statistic called **student's *t*-test** is used. The *t* statistic is calculated as follows:

$$t = \frac{\overline{X}_1 - \overline{X}_2}{s_p \sqrt{1/n_1 + 1/n_2}}$$

26-1

where

$\overline{X}_1$ = arithmetic mean of Treatment 1
$\overline{X}_2$ = arithmetic mean of Treatment 2
n_1 = number of samples in Treatment 1
n_2 = number of samples in Treatment 2
s_p = pooled variance of the samples

$$s_p = \frac{\sqrt{(n_1 - 1)s_1^2 + (n_2 - 1)s_2^2}}{\sqrt{n_1 + n_2 - 2}}$$

26-2

Once the *t* statistic has been calculated for a comparison, it is compared to a table of critical *t* values found in most statistics textbooks. You find the appropriate degrees of freedom for your comparison ($n_1 + n_2 - 2$ in this case, the degrees of freedom are always one less than the number of samples) and read across to the column that shows the level of significance you have chosen. If the calculated *t* value is greater than the tabular *t* value, then any difference in the means of the two treatments is assumed to be statistically significant.

As you can see, the number crunching starts to become significant even with simple statistics, hence, the value of packaged computer statistics programs. Microsoft Excel has some simple statistics that will automatically calculate the t statistic for you once the data are entered. Hints on how to use the statistical package in Microsoft Excel are in Appendix 11.

Example 26–1

Is the mean N content in Soil A different from Soil B given that five replicated measurements from each soil gave the following N values (mg kg^{-1}):

$$\text{Soil A—18, 16, 18, 15, 19}$$

$$\text{Soil B—17, 17, 18, 14, 19}$$

Solution

The null hypothesis (H_0) is that N content in Soil A = Soil B. Let's assume a significance level of $\alpha = 0.05$ (5 percent probability that the null hypothesis will be falsely rejected). First, organize the data to make calculating the statistics easier.

Sample	Soil A		Soil B	
	X_A	X_A^2	X_B	X_B^2
1	18	324	17	289
2	16	256	17	289
3	18	324	18	324
4	15	225	14	196
5	19	361	19	361
	$\sum X_A = 86$	$\sum X_A^2 = 1490$	$\sum X_B = 85$	$\sum X_B^2 = 1459$

$$\overline{X}_A = 86/5 = 17.2$$

$$\overline{X}_B = 85/5 = 17.0$$

The variance for each sample is calculated as you have seen in Chapter 24.

$$s_A^2 = \frac{\sum X_A^2 - \frac{\left(\sum X_A\right)^2}{n}}{n-1} \qquad s_B^2 = \frac{\sum X_B^2 - \frac{\left(\sum X_B\right)^2}{n}}{n-1}$$

$$= \frac{1490 - \frac{(86)^2}{5}}{5-1} \qquad\qquad = \frac{1459 - \frac{(85)^2}{5}}{5-1}$$

$$= 2.7 \qquad\qquad\qquad\qquad = 3.5$$

The pooled variance (s_p) = $\dfrac{\sqrt{(n_A - 1)s_A^2 + (n_B - 1)s_B^2}}{\sqrt{n_A + n_B - 2}}$

$$= \dfrac{\sqrt{(5 - 1)2.7 + (5 - 1)3.5}}{\sqrt{5 + 5 - 2}}$$

$$= 1.76$$

Putting it all together gives

$$t = \dfrac{\overline{X}_A - \overline{X}_B}{s_p \sqrt{1/n_1 + 1/n_2}}$$

$$t = \dfrac{17.2 - 17.0}{1.76 \sqrt{1/5 + 1/5}}$$

$$t = 0.18$$

If you look at Table 26–2, and read across from 8 degrees of freedom to the column under $\alpha = 0.05$ (one-tailed test), the tabulated t statistic is 1.860. Since $0.18 < 1.860$, you can safely accept the null hypothesis that N content is the same in Soil A and Soil B.

The problem above was an example of a one-tailed t-test in which the only alternative hypothesis was that the N content in Soil A was greater than in Soil B (Soil A > Soil B). If you didn't know anything about the sites in question, an equally valid alternative hypothesis would be that the N content was simply not the same in Soil A and Soil B (Soil A $\neq$ Soil B). Either Soil A > Soil B or Soil A < Soil B. This would have been an example of a two-tailed t-test. The method of calculating student's t statistic is exactly the same. However, the tabulated t statistic is slightly different for each level of significance (Table 26–2).

Example 26–2

The t statistic calculated for two treatment means was 1.10. Are these treatment means the same at a significance level of 0.05 if there were 12 degrees of freedom and the alternative hypothesis was that the two treatment means were not equal?

Solution

This is an example of a two-tailed t-test. The significance level is 0.05, the calculated t statistic is 1.10. There are 12 degrees of freedom. From Table 26–2 you can see that the tabulated t statistic for these conditions is 2.179. Since calculated $t <$ tabulated t, these treatment means are not significantly different. You can reject the alternative hypothesis.

The t-test can also be used to compare a set of measurements with a theoretical value or a standard value, which, for example, you might obtain from a

TABLE 26–2   Critical Values of Student's *t* for One- and Two-Tailed Tests

Degrees of Freedom (df)	One-tailed ($X > \mu$) Two-tailed ($X \neq \mu$)	0.25 0.50	0.05 0.10	0.025 0.05	0.005 0.01	0.0005 0.001
1		1.000	6.314	12.706	63.657	636.619
2		0.816	2.920	4.303	9.925	31.598
3		0.765	2.353	3.182	5.841	12.941
4		0.741	2.132	2.776	4.604	8.610
5		0.727	2.015	2.571	4.032	6.859
6		0.718	1.943	2.447	3.707	5.959
7		0.711	1.895	2.365	3.499	5.405
8		0.706	1.860	2.306	3.355	5.041
9		0.703	1.833	2.262	3.250	4.781
10		0.700	1.812	2.228	3.169	4.587
11		0.697	1.796	2.201	2.106	4.437
12		0.695	1.782	2.179	3.055	4.318
13		0.694	1.771	2.160	3.012	4.221
14		0.692	1.761	2.145	2.977	4.140
15		0.691	1.753	2.131	2.947	4.073
16		0.690	1.746	2.120	2.921	4.015
17		0.689	1.740	2.110	2.898	3.965
18		0.688	1.734	2.101	2.878	3.922
19		0.688	1.729	2.093	2.861	3.883
20		0.687	1.725	2.086	2.845	3.850

Note. Steele, R. G., & Torrie, J. H. (1980). *Principles and procedures of statistics: A biometrical approach* (2nd ed.). New York: McGraw-Hill.

standard laboratory procedure. For this type of comparison, the following equation is used:

$$|t| = \frac{\overline{X} - \mu}{s/\sqrt{n}}$$ 26-3

where

t = calculated t statistic
$\overline{X}$ = arithmetic mean of the sample data
μ = standard value or theoretical value
s = standard deviation of the sample data
n = number of measurements

If the absolute value of the t statistic (the numerical value of the t statistic with its sign ignored) is greater than the tabulated t statistic, then, at the chosen level of significance, there is a significant difference between the measures and standard means.

Example 26–3

Seven measurements of soil respiration were made at a field site giving values of 0.96, 1.04, 0.95, 1.0, 1.02, 1.11, and 0.98 kg CO_2 m^{-2} year^{-1}. Is the mean of these values significantly different (at $\alpha = 0.01$) from a theoretical value of 0.26 kg CO_2 m^{-2} year^{-1}?

Solution

There are seven measurements and only one data set is used, so the degrees of freedom in this example are $n - 1$ or $7 - 1 = 6$. $\overline{X}$ of the data is 1.01, $s = 0.05$, $n = 7$, and $\mu = 0.26$. Plugging these values into the appropriate equation for the t statistic gives

$$t = \frac{1.01 - 0.26}{0.05/\sqrt{7}} = \frac{0.75}{0.02} = 37.50$$

From Table 26–2, the tabulated t statistic at $\alpha = 0.01$ and 6 degrees of freedom is 3.707. Since $37.50 > 3.707$, the sample mean is significantly different from the theoretical mean.

Linear Regression

In many instances in soil science, two variables are directly related to one another. For example, the relationship between NO_3^- content in soil and denitrification. This relationship is often linear, that is, the magnitude of the dependent variable (the variable that changes in response to another) is proportional to the magnitude of the independent variable. The independent variable is usually denoted as X and the dependent variable as Y.

Plotting X versus Y usually doesn't give a straight line if you connect the data points. So, you need to calculate the **best fit** of this data. That is, you need to calculate the slope ($\Delta Y/\Delta X$) and Y intercept of a line that generates the smallest deviation of all the data points from the line. The method used to calculate this line is the **least squares formula.**

Example 26–4

Calculate the linear regression of the following data using a least squares best fit:

$$X = 1.00, 0.25, 0.50, 0.65$$

$$Y = 1.00, 0.27, 0.57, 0.85$$

Solution

First, calculate the mean values and sums of squares for X and Y.

X	Y	$(X)(Y)$
0.25	0.27	0.07
0.50	0.57	0.29
0.65	0.85	0.55
1.00	1.00	1.00
$\sum X = 2.40$	$\sum Y = 2.69$	$\sum XY = 1.91$

$$n = 4 \qquad\qquad n = 4$$

$$\overline{X} = 0.6 \qquad\qquad \overline{Y} = 0.67$$

$$\sum X^2 = 1.74 \qquad \sum Y^2 = 2.12$$

Calculate the squared deviations $[(\sum X - \overline{X})^2]$ for the means of the independent variable X. Identify this value as $\sum x^2$.

$$\sum x^2 = \sum X^2 - \frac{(\sum X)^2}{n} = 1.74 - \frac{(2.4)^2}{4} = 0.30$$

Calculate the cross products of the deviations from the mean.

$$\sum xy = \sum (X - \overline{X})(Y - \overline{Y})$$

$$\sum xy = \sum XY - \frac{(\sum X)(\sum Y)}{n}$$

$$\sum xy = 1.91 - \frac{(2.4)(2.69)}{4}$$

$$\sum xy = 0.3$$

The slope (m) of a line that gives the least squares best fit through the data is

$$m = \frac{\sum xy}{\sum x^2} = 0.30/0.30 = 1.0$$

Since the equation for a line is $y = mx \pm b$, you can solve for the Y intercept (b) by plugging in the average values of X and Y and the calculated slope (m).

$$Y = mX \pm b$$

$$Y - mX = \pm b$$

$$0.67 - (1.0)(0.6) = \pm b$$

$$0.07 = b$$

This is all the information you need to know to plot the linear regression. The equation of the line is

$$Y = mX \pm b \quad \text{or} \quad Y = (1.0)(X) + 0.07$$

If you know any value of X, you should be able to estimate the corresponding value of Y.

The calculation of the linear regression was extremely tedious. Fortunately, most scientific calculators have a program for calculating linear regression that requires you to input only the X and Y variables. Still, it's nice to go through the process just once to see what happens to all those numbers.

Correlation Coefficients

Although you were able to calculate a linear regression using a best fit method, it would be nice to know just how linear the data really are. To do this, a statistic called the **correlation coefficient** (r) is used. The r ranges from +1 to −1. The closer the r value is to either extreme, the more linear the data. If r is 0, or close to 0, there is no relationship between the two variables (at least not a linear one).

To calculate r, use the following equation:

$$r = \frac{\sum xy}{\sqrt{\sum x^2 \sum y^2}}$$

26-4

Example 26–5

Calculate the correlation coefficient for the data in Example 26–4.

Solution

$$\sum X = 2.40 \quad \text{and} \quad \sum Y = 2.69$$

$$\sum xy = \sum XY - \frac{\sum X \sum Y}{n}$$

$$\sum xy = 1.91 - \frac{(2.40)(2.69)}{4} = 0.30$$

The sum of squared deviations $[(X - \bar{X})^2 \text{ or } (Y - \bar{Y})^2]$ for X and Y, respectively, is

$$\sum x^2 = \sum X^2 - \frac{(\sum X)^2}{n} \qquad \sum y^2 = \sum Y^2 - \frac{(\sum Y)^2}{n}$$

$$\sum x^2 = 1.74 - \frac{(2.4)^2}{4} \qquad \sum y^2 = 2.12 - \frac{(2.69)^2}{4}$$

$$\sum x^2 = 0.30 \qquad \sum y^2 = 0.31$$

Plugging all these values into the equation for the correlation coefficient gives

$$r = \frac{0.30}{\sqrt{(0.30)(0.31)}} = 0.98$$

Once more, this is a tedious calculation. But, as with linear regressions, most scientific calculators have this statistical function already programmed.

Using t Statistics to Test the Probability of Linearity

After you calculate the linear regression and a correlation coefficient to assess just how linear your data is, it would still be nice to determine the probability that the data are indeed linear. You can use student's t statistic to calculate this with the following equation:

$$t = \frac{r}{s_r}$$

26-5

where

r = correlation coefficient
s_r = standard error

s_r is calculated as

$$s_r = \sqrt{\frac{(1 - r^2)}{(n - 2)}}$$

26-6

Example 26–6

For the data in Example 26–4, determine whether the data are significantly linear at $\alpha = 0.05$.

Solution

$r = 0.98$ as previously calculated

$r^2 = 0.96$

$n = 4$

$$t = \frac{0.98}{\sqrt{(1 - 0.96)/(4 - 2)}}$$

$$= \frac{0.98}{0.14}$$

$$= 7.00$$

Since there are two data sets, the degrees of freedom = $n - 2$ or 2. This is a two-tailed test, so at $\alpha = 0.05$, the tabulated t statistic is 4.303. Since $7.00 > 4.303$ the data are considered to be significantly linear.

Confidence Limits

The mean value of a data set is really only an estimate of what the actual value might be. Confidence limits can be calculated for the data set to indicate the precision of the estimated mean. To calculate confidence limits, you need to calculate the standard error of the data set. The standard error is simply

$$s_x = \frac{s}{\sqrt{n}}$$ **26-7**

Remember that the working formula for calculating standard deviations is

$$s^2 = \frac{\sum X^2 - \frac{(\sum X)^2}{n}}{n-1} \qquad s = \sqrt{s^2}$$

If you know Xs_x, and the degrees of freedom (df), you can calculate confidence limits.

Example 26–7

Calculate the 95 percent confidence limits for the following data set of BOD measurements (mg L^{-1}): 110, 92, 84, 119, 101, 81, 129, 136, 126, 86.

Solution

The mean value ($\overline{X}$) is 106 mg L^{-1}. The standard error is $s/\sqrt{n}$. The standard deviation is equal to 20, and the number of measurements (n) = 10, so $s_x = 20/\sqrt{10} = 6$. The degrees of freedom for each data set you use are always one less than the number of observations in the data set. So, the degrees of freedom in this example are $10 - 1 = 9$.

The critical value of t for this example can be looked up in Table 26–2. This is a two-tailed test with 9 degrees of freedom and a μ value of 0.05. So, the critical t statistic is 2.262. The confidence limit is given by the equation

$$\overline{X} \pm t\,(s_x)$$ **26-8**

For this example the corresponding values are

$$106 + 2.262\,(6) = 120 \text{ mg L}^{-1}$$
$$106 - 2.262\,(6) = 92 \text{ mg L}^{-1}$$

Variance Analysis

The F test is a method of testing whether the variance of two data sets is comparable. This is important if you want to compare the significance, say, of soil NO_3^- concentrations measured by two different methods. The equation for the F-test is quite simple.

$$F = \frac{s_1^2}{s_2^2}$$ **26-9**

where

F = F-ratio value
s_1^2 = variance of the first data set (always the numerically larger number so that the F-ratio is >1)
s_2^2 = variance of the second data set

You already know how to calculate the variance of data sets; use your calculator's programmed functions or this working formula:

$$s^2 = \frac{\sum X^2 - \frac{(\sum X)^2}{n}}{n - 1}$$

Once this calculation is completed, determine the degrees of freedom for each data set. The degrees of freedom are always one less than the number of observations. Finally, look up a table of tabulated F values (Table 26–3 gives an example) and locate the intersection of the column and row that match the numerator and denominator. At that intersection, find the level of significance you're using. If your calculated F-ratio is less than this tabulated F-ratio, then the two variances are not significantly different.

TABLE 26–3 Critical Values of F for One- and Two-Tailed Tests

Denominator df	Probability of a Larger F		Numerator df				
	1-tailed	2-tailed	1	2	3	4	5
1	.05	.10	161.4	199.5	215.7	244.6	230.2
	.025	.05	647.8	799.5	864.2	899.6	921.8
	.005	.01	16,211	20,000	21,615	22,500	23,056
2	.05	.10	18.5	19.0	19.2	19.2	19.3
	.025	.05	38.5	39.0	39.2	39.2	39.3
	.005	.01	198.5	199.0	194.2	199.2	199.3
3	.05	.10	10.1	9.6	9.3	9.1	9.0
	.025	.05	17.4	16.0	15.4	15.1	14.9
	.005	.01	55.6	49.8	47.5	46.2	45.4
4	.05	.10	7.7	6.9	6.6	6.4	6.3
	.025	.05	12.2	10.6	10.0	9.6	9.4
	.005	.01	31.3	26.3	24.3	23.2	22.5
5	.05	.10	6.6	5.8	5.4	5.2	5.0
	.025	.05	10.0	8.4	7.8	7.4	7.2
	.005	.01	22.8	18.3	16.5	15.6	14.9

Note. Steele, R. G., & Torrie, J. H. (1980). *Principles and procedures of statistics: A biometrical approach,* (2nd ed.). New York: McGraw-Hill.

Example 26–8

Compare the variance of two methods to measure NH_4^+ concentration in soil samples. Are the methods statistically different at $\alpha = 0.25$?

Method 1. (ppm NH_4^+) : 17, 18, 19, 18, 16
Method 2. (ppm NH_4^+): 18, 18, 19, 19

Solution

Calculate the variance of each method.

Method 1	Method 2
17	18
18	18
19	19
18	19
16	
$\Sigma X_1 = 88$	$\Sigma X_2 = 74$

$$\Sigma X_1^2 = 1554 \qquad \Sigma X_2^2 = 1370$$

$$\overline{X}_1 = 18 \qquad \overline{X}_2 = 19$$

$$n_1 = 5 \qquad n_2 = 4$$

$$s_1 = 1.14 \qquad s_2 = 0.58$$

$$s_1^2 = 1.30 \qquad s_2^2 = 0.33$$

$$F = \frac{s_1^2}{s_2^2} = \frac{1.30}{0.33} = 3.94$$

Table 26–3 shows that for a numerator with 4 degrees of freedom and a denominator with 3 degrees of freedom and an α value of 0.25, the tabulated F value is 15.10. Since $3.94 < 15.10$, there is no significant difference between the two methods.

Analysis of variance (ANOVA) is one of the principle statistical methods used to evaluate the statistical significance between samples that have multiple treatments imposed on them. However, a discussion of that topic is beyond the scope of this workbook and you should refer to introductory statistics textbooks for this information.

References

Atlas, R. M., & Bartha, R. (1993). *Microbial ecology: Fundamentals and applications* (3rd ed.). New York: Benjamin.

Carmer, S. G., & Walker, W. M. (1988). Significance from a statistician's viewpoint. *Journal of Production Agriculture, 1,* 27–33.

Kiazolu, J. S., Mrema, M. J., & Sebolai, B. (1994). *A guide to agricultural research methodology and report writing for senior secondary school teachers in Botswana.* Ministry of Education. Government of Botswana.

Masterson, W. L., & Slowinski, E. L. (1970). *Mathematical preparation for general chemistry.* Philadelphia, PA: W. B. Saunders.

Rees, D. G. (1989). *Essential statistics.* New York: Chapman & Hall.

Steele, R. G., & Torrie, J. H. (1980). *Principles and procedures of statistics: A biometrical approach* (2nd ed.). New York: McGraw-Hill.

Sample Problems

Answer the following questions using the data set shown below for a study examining the effect of manure on corn yield.

Replication	Manure added (Trt$_A$)	No manure (Trt$_B$)
	Yield (Mg ha^{-1})	
1	5.0	1.2
2	4.2	1.0
3	4.6	1.8
4	4.8	1.9
5	4.4	1.5

1. Use a *t*-test to test the hypothesis that adding manure significantly increased corn yield compared to unmanured treatments. Use a significance level of 0.05.
2. The national mean corn yield for this country in unmanured plots is 2.0 Mg ha^{-1}. Is the yield from Treatment B significantly different from the national yield at a significance level of 0.05?
3. What are the 99.9 percent confidence limits for Trt$_A$ and Trt$_B$?
4. Is the variance of Trt$_A$ and Trt$_B$ significantly different at $\alpha = 0.05$?

Answer the following questions using the data sets shown below for a study examining the effect of temperature on C mineralization in soil:

Temperature (°C)	C Mineralized (mg kg^{-1})
5.2	14
4.8	14
7.2	21
8.9	26
8.0	26
13.0	40
13.9	40
17.3	44
20.8	52
22.5	53
25.9	65
19.9	66
25.5	68

1. Use the least squares method to find the best fit for a linear regression through these data.
2. Determine the correlation coefficient for these data.
3. Determine whether C mineralized and temperature are significantly correlated at $\alpha = 0.05$.

27

Sampling

OBJECTIVE

The objective of this chapter is to introduce

- methods of sampling.
- objective determination of the number of samples to take.

Overview

You can't analyze the entire soil. If it were not already nearly impossible, it would be prohibitively expensive, and if neither cost nor labor were an issue, analyzing the entire soil would defeat the purpose of subsequently managing it—the soil would be in the lab. Soil scientists are forced to use samples, which unfortunately, are subject to the heterogeneity of the soil. Two samples from different locations may or may not resemble one another. The questions this chapter addresses are how to sample the soil in a representative manner, and how to objectively determine the number of samples that are necessary to reliably represent soil properties.

Sampling Schemes

The difficulty in soil sampling is deciding where and when to sample. Although some soil properties such as %C and texture are essentially constant, except in the long term, other properties such as the NO_3^- concentration or the microbial biomass can fluctuate on a seasonal and sometimes on a

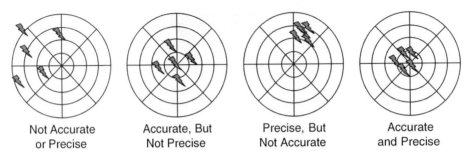

Not Accurate or Precise	Accurate, But Not Precise	Precise, But Not Accurate	Accurate and Precise

FIGURE 27–1   The difference between precision and accuracy illustrated [*adapted from* HACH. (1997). *HACH water analysis handbook* (3rd ed.). Loveland, CO: Author].

daily basis. The soil scientist aims for precision and hopes for accuracy. What's the difference between the two? **Precision** reflects how close measurements are to one another—reproducibility. **Accuracy** reflects how close one is to the "right" value. Figure 27–1 illustrates the difference between precision and accuracy.

Sampling schemes are generally designed to provide the greatest precision for a fixed cost or some specified precision at a lower cost (Wollum, 1994). There are four basic sampling schemes—**judgmental, simple random, stratified random, and systematic.** No scheme is ideal for every soil system. Judgmental sampling is an openly biased sampling scheme in which samples are only taken from areas that are judged to be representative (Figure 27–2). So, the samples aren't really representative at all, because they don't reflect the soil's heterogeneity.

In simple random sampling, no location in a specified site is more or less likely to be selected for sampling than any other. It is usually best to randomly pick the sampling sites before going to the field, otherwise some intentional bias could occur. The danger of this scheme is that some portions of the site may not be sampled through chance alone.

Stratified random sampling applies the principles of random sampling to subareas of the site that are suspected to be different. This strategy is defeated if too many subareas are designated.

Systematic sampling, or grid sampling, ensures that the entire site is represented by individual samples. Each sample site is predetermined at uniform intervals. Figure 27–2 shows this sampling procedure as a pair of transects. In some cases, these individual samples can be treated as though they were randomly selected. The independence of each sample would be suspect, however, if the transects shown in Figure 27–2 both followed the soil slope. Precision agriculture is highly dependent on grid sampling to develop management recommendations. One of the big issues facing precision agriculture is how much distance there can be between points on the grid that will still allow you to precision manage.

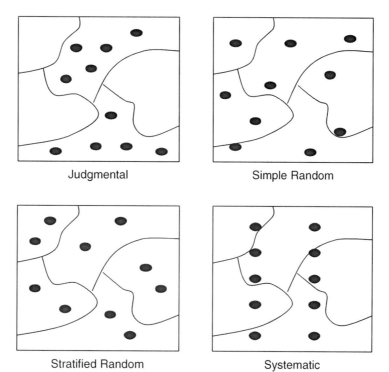

FIGURE 27–2   Four sampling schemes illustrating how to sample a location. Each dot represents a different sample point and each sector represents a different soil series or topographic environment.

Compositing Samples

Soil samples are frequently composited to reduce the number of samples that have to be analyzed. Certain conditions have to be met to make this an acceptable procedure (Wollum, 1994).

1. The composites should consist of an equal number and amount of each sampling unit.
2. The sampling units must not interact (one unit must not have been influenced by another unit).
3. The only objective is to estimate the mean of some property from the composite.
4. One can obtain an estimate of variance between similar composite samples, but not for either the original site or individual sampling units.

Estimating the Number of Samples to Take

Based on a literature survey or some preliminary sampling, you may have a feel for the mean and variance of a soil property. With that information you may be able to predict the number of samples you have to take to obtain a certain level of precision (perhaps to be within 10 percent of the true mean) with a certain level of confidence. The equation to use is as follows:

$$\text{No. of samples} = \frac{(t \text{ value})^2(\text{sample variance})}{[(\text{estimated sample mean})(\text{level of precision})]^2} \qquad \boxed{\textbf{27-1}}$$

Example 27–1

A previous sample of a soil based on 10 samples suggests that the mean NO_3^- concentration is 50 ppm with a variance of 25 ppm. How many samples do you have to take to be 99 percent sure of being within 5 percent of the mean?

Solution

The t value for 99 percent confidence and $10 - 1(n - 1)$ degrees of freedom is 3.250, the variance is 25 ppm, the estimated sample mean is 50 ppm, and the level of precision is 5 percent or 0.05. Plugging these values into our equation gives

$$\text{No. of samples} = \frac{(3.250)^2\,(25)}{[(50)(0.05)]^2} = 42$$

If one were less fastidious about the precision of the measurement (10 percent rather than 5 percent) or less stringent about the confidence level (95 percent rather than 99 percent) the required number of samples would decrease to 5.

Reference

Wollum, A. G. (1994). Soil sampling for microbiological analysis. In R. W. Weaver, et al. (ed.) *Methods of soil analysis. Part 2. Microbiological and bio-chemical properties* (pp. 1–14). Madison, WI: Soil Science Society of America.

Sample Problems

1. The mean sand content of a site based on 8 samples was estimated to be 15 percent with a coefficient of variation (CV) of 20 percent. How many samples would you have to take from this site to be 95 percent confident that you had estimated the true mean with a precision of ±10 percent?

2. The mean denitrification rate in a uniform site is estimated to be 50 mg N m^{-2} day^{-1} with a CV of 200 percent based on 10 samples. How many samples would you need to take to estimate the true mean within 20 percent at a confidence of 99 percent?

3. In Question 2 above, how many samples could you omit if you were content with a confidence level of 10 percent?

4. The measured gravimetric water content of a site is 30 percent with a CV of 100 percent based on two samples. Which is less work to estimate the true mean: resample eight times to reduce the CV to 10 percent or take the required number of calculated samples, assuming that you want to be within 5 percent of the true mean with a confidence level of 95 percent?

Appendices

pK_a Values of Common Environmentally Important Buffering Systems

Buffer System	pK_a
$H_2SO_4 \Leftrightarrow H^+ + HSO_4^-$	2.0
$H_3PO_4 \Leftrightarrow H^+ + H_2PO_4^-$	2.23
Citric acid $\Leftrightarrow H^+ +$ citrate monobasic	3.13
Citrate monobasic $\Leftrightarrow H^+ +$ citrate dibasic	4.76
Citrate dibasic $\Leftrightarrow H^+ +$ citrate tribasic	6.39
$H_2CO_3 \Leftrightarrow H^+ + HCO_3^-$	6.4
$H_2S \Leftrightarrow H^+ + HS^-$	7.0
$H_2PO_4^- \Leftrightarrow H^+ + H_2PO_4^{2-}$	7.2
Tris acid $\Leftrightarrow H^+ +$ tris base	8.08
$HCN \Leftrightarrow H^+ + CN^-$	9.2
$NH_4^+ \Leftrightarrow H^+ + NH_3$	9.2
$H_2BO_3 \Leftrightarrow H^+ + HBO_3^-$	9.2
$H_2SiO_4 \Leftrightarrow H^+ + HSiO_4^-$	9.5
NH_2CH_2COOH (glycine) $\Leftrightarrow H^+ + NH_2CH_2COO^-$	9.87
$HCO_3^{3-} \Leftrightarrow H^+ + CO_3^{2-}$	10.33
$HPO_4^{2-} \Leftrightarrow H^+ + PO_4^{3-}$	12.3
$HS^- \Leftrightarrow H^+ + S^{2-}$	12.9
$H_2O \Leftrightarrow H^+ + OH^-$	14.0

Segel, I. H. (1976). *Biochemical calculations* (2nd ed.). New York: John Wiley & Sons.

2

Formulas for Stock Solutions and Buffers of Various pH (The Buffer Cheat Sheet)

1. HCL-KCL Buffer

Stock Solutions

A. 0.2 M KCL (14.91 g in 1000 mL)
B. 0.2 M HCL (16.5 mL of conc. HCL in 1000 mL)

Formula: 50 mL A + ? mL B diluted to 200 mL

pH	1.0	1.1	1.2	1.3	1.4	1.5	1.6
?	97.0	78.0	64.5	51.0	41.5	33.3	26.3

pH	1.7	1.8	1.9	2.0	2.1	2.2
?	20.6	16.6	13.2	10.6	8.4	6.7

2. HCL-Glycine Buffer
Stock Solutions
A. 0.2 M glycine solution (15.01 g in 1000 mL)
B. 0.2 M HCL (16.5 mL of conc. HCL in 1000 mL)

Formula: 50 mL A + ? mL B diluted to 200 mL

pH	2.2	2.4	2.6	2.8	3.0	3.2	3.4	3.6
?	44.0	32.4	24.2	18.8	11.4	8.2	6.4	5.0

3. HCL-Phthalate Buffer
Stock Solutions
A. 0.2 M potassium acid phthalate solution (40.84 g in 1000 mL)
B. 0.2 M HCL (16.5 mL of conc. HCL in 1000 mL)

Formula: 50 mL A + ? mL B diluted to 200 mL

pH	2.2	2.4	2.6	2.8	3.0	3.2	3.4	3.6	3.8
?	46.7	39.6	33.0	26.4	20.3	14.7	9.9	6.0	2.63

4. Aconitate Buffer
Stock Solutions
A. 0.5 M aconitate solution (87.05 g in 1000 mL)
B. 0.2 M NaOH (8 g NaOH in 1000 mL)

Formula: 20 mL A + ? mL B diluted to 200 mL

pH	2.5	2.7	2.9	3.1	3.3	3.5	3.7	3.9	4.1	4.3
?	15.0	21.0	28.0	36.0	44.0	52.0	60.0	68.0	76.0	83.0

pH	4.5	4.7	4.9	5.1	5.3	5.5	5.7
?	90.0	97.0	103	108	113	119	126

5. Citrate Buffer

Stock Solutions

A. 0.1 M citric acid (21.01 g in 1000 mL)
B. 0.1 M sodium citrate (29.41 g $C_6H_5O_7Na_3 \cdot 2H_2O$ in 1000 mL)(do not use $5\frac{1}{2}$ water)

Formula: x mL A + y mL B diluted to 100 mL

pH	3.0	3.2	3.4	3.6	3.8	4.0	4.2	4.4	4.6	4.8
x	46.5	43.7	40.0	37.0	35.0	33.0	31.5	28.0	25.5	23.0
y	3.5	6.3	10.0	13.0	15.0	17.0	18.5	22.0	24.5	27.0

pH	5.0	5.2	5.4	5.6	5.8	6.0	6.2
x	20.5	18.0	16.0	13.7	11.8	9.5	7.2
y	29.5	32.0	34.0	36.3	38.2	41.5	42.8

6. Acetate Buffer

Stock Solutions

A. 0.2 M acetic acid (11.55 mL in 1000 mL)
B. 0.1 M sodium acetate (27.2 g $C_2H_3O_2Na \cdot 3H_2O$ in 1000 mL)
(or 16.4 g $C_2H_3O_2Na$ in 1000 mL)

Formula: x mL A + y mL B diluted to 100 mL

pH	3.6	3.8	4.0	4.2	4.4	4.6	4.8	5.0	5.2	5.4	5.6
x	46.3	44.0	41.0	36.8	30.5	25.5	20.0	14.8	10.5	8.8	4.8
y	3.7	6.0	9.0	13.2	19.5	24.5	30.0	35.2	39.5	41.2	45.2

7. Citrate-Phosphate Buffer

Stock Solutions

A. 0.1 M citric acid (19.21 mL in 1000 mL)
B. 0.2 M dibasic NaPO$_4$ (53.65 g Na$_2$HPO$_4$•7H$_2$O or 71.7 Na$_2$HPO$_4$•12H$_2$O in 1000 mL)

Formula: x mL A + y mL B diluted to 100 mL

pH	2.6	2.8	3.0	3.2	3.4	3.6	3.8	4.0	4.2	4.4	4.6	4.8
x	44.6	42.2	39.8	37.7	35.9	33.9	32.3	30.7	29.4	27.8	26.7	25.2
y	5.4	7.9	10.2	12.3	14.1	16.1	17.7	19.3	20.6	22.2	23.3	24.8

pH	5.0	5.2	5.4	5.6	5.8	6.0	6.2	6.4	6.6	6.8	7.0	
x	24.3	23.3	22.2	21.0	19.7	17.9	16.9	15.4	13.6	9.1	6.5	
y	25.7	27.8	29.0	30.3	30.3	32.1	33.1	34.6	36.4	40.9	43.6	

8. Succinate Buffer

Stock Solutions

A. 0.2 M succinate solution (23.6 g in 1000 mL)
B. 0.2 M NaOH (8 g NaOH in 1000 mL)

Formula: 25 mL A + ? mL B diluted to 100 mL

pH	3.8	4.0	4.2	4.4	4.6	4.8	5.0	5.2	5.4	5.6	5.8	6.0
?	7.5	10.0	13.3	16.7	20.0	23.5	26.7	30.3	34.2	37.5	40.7	43.5

9. Phthalate-NaOH Buffer

Stock Solutions

A. 0.2 M K acid phthalate (40.84 g in 1000 mL)
B. 0.2 M NaOH (8 g NaOH in 1000 mL)

Formula: 50 mL A + ? mL B diluted to 200 mL

pH	4.2	4.4	4.6	4.8	5.0	5.2	5.4	5.6	5.8	6.0
?	3.7	7.5	12.2	17.7	23.9	30.0	35.5	39.8	43.0	45.5

10. Maleate Buffer

Stock Solutions

A. 0.2 M acid Na maleate (8 g NaOH + 23.2 g maleic acid or 19.6 g maleic anhydride in 1000 mL)
B. 0.2 M NaOH (8 g NaOH in 1000 mL)

Formula: 50 mL A + ? mL B diluted to 200 mL

pH	5.2	5.4	5.6	5.8	6.0	6.2	6.4	6.6	6.8
?	7.2	10.5	15.3	20.8	26.9	33.0	38.0	41.6	44.4

11. Cacodylate Buffer

Stock Solutions

A. 0.2 M Na cacodylate (42.8 g of $Na(CH_3)_2AsO_2 \cdot 3H_2O$ in 1000 mL)
B. 0.2 M HCL (16.5 mL of conc. HCL in 1000 mL)

Formula: 50 mL A + ? mL B diluted to 200 mL

pH	5.0	5.2	5.3	5.6	5.8	6.0	6.2
?	47.0	45.0	43.0	39.2	34.8	29.6	23.8

pH	6.4	6.6	6.8	7.0	7.2	7.4
?	18.3	13.3	9.3	6.3	4.2	2.7

12. Phosphate Buffer

Stock Solutions

A. 0.2 M NaH_2PO_4 (27.8 g in 1000 mL) or KH_2PO_4 (27.2 g in 1000 mL)
B. 0.2 M Na_2HPO_4 (53.65 g $Na_2HPO_4 \cdot 7H_2O$ or 71.7, $Na_2HPO_4 \cdot 12H_2O$ in 1000 mL) or K_2HPO_4 (34.8 g in 1000 mL)

Formula: x mL A + y mL B diluted to 100 mL (diluted to 1000 mL to give 0.02 M buffer

pH	5.7	5.8	5.9	6.0	6.1	6.2	6.3	6.4
x	93.5	92.0	90.0	87.7	85.0	81.5	77.5	73.5
y	6.5	8.0	10.0	12.3	15.0	18.5	22.5	26.5

pH	6.5	6.6	6.7	6.8	6.9	7.0	7.1	7.2
x	68.5	62.5	56.5	51.0	45.0	39.0	33.0	28.0
y	31.5	37.5	43.5	49.0	55.0	61.0	67.0	72.0

pH	7.3	7.4	7.5	7.6	7.7	7.8	7.9	8.0
x	23.0	19.0	16.0	13.0	10.5	8.5	7.0	5.3
y	77.0	81.0	84.0	87.0	90.5	91.5	93.0	94.7

13. Tris(hydroxymethyl)aminomethane-Maleate Buffer

Stock Solutions

A. 0.2 M solution of tris acid maleate (24.2 g tris(hydroxymethyl) aminomethane + 23.2 g maleic acid or 19.6 g maleic anhydride in 1000 mL)

B. 0.2 M NaOH (8.0 g NaOH in 1000 mL)

Formula: 50 mL A + ? mL B diluted to 200 mL

pH	5.2	5.4	5.6	5.8	6.0	6.2	6.4	6.6	6.8
?	7.0	10.8	15.5	20.5	26.0	31.5	37.0	42.5	45.0

pH	7.0	7.2	7.4	7.6	7.8	8.0	8.2	8.4	8.6
?	48.0	51.0	54.0	58.0	63.5	69.0	75.0	81.0	86.5

14. Barbital (Veronal) Buffer

Stock Solutions

A. 0.2 M Na barbital (veronal) (41.2 g in 1000 mL)

B. 0.2 M HCL (16.5 mL of conc. HCL in 1000 mL)

Formula: 50 mL A + ? mL B diluted to 200 mL

pH	6.8	7.0	7.2	7.4	7.6	7.8	8.0
?	45.0	43.0	39.0	32.5	27.5	22.5	17.5

pH	8.2	8.4	8.6	8.8	9.0	9.2
?	12.7	9.0	6.0	4.0	2.5	1.5

15. Tris(hydroxymethyl)aminomethane Buffer

Stock Solutions

A. 0.2 M tris(hydroxymethyl)aminomethane (24.2 g in 1000 mL)

B. 0.2 M HCL (16.5 mL of conc. HCL in 1000 mL)

Formula: 50 mL A + ? mL B diluted to 200 mL

pH	7.2	7.4	7.6	7.8	8.0	8.2	8.4	8.6	8.8	9.0
?	44.2	41.4	38.4	32.5	26.8	21.9	16.5	12.2	8.1	5.0

16. Boric Acid-Borax Buffer

Stock Solutions

A. 0.2 M boric acid (12.4 g in 1000 mL)
B. 0.05 M borax solution (19.5 g in 1000 mL; 0.2 M in terms of Na borate)

Formula: 50 mL A + ? mL B diluted to 200 mL

pH	7.6	7.8	8.0	8.2	8.4	8.6
?	2.0	3.1	4.9	7.3	11.5	17.5

pH	8.7	8.8	8.9	9.0	9.1	9.2
?	22.5	30.0	42.5	59.0	83.0	115.0

17. 2-amino-2-methyl-1,3-propanediol (Ammediol) Buffer

Stock Solutions

A. 0.2 M solution of 2-amino-2-methyl-1,3-propanediol (21.03 g in 1000 mL)
B. 0.2 M HCL (16.5 mL of conc. HCL in 1000 mL)

Formula: 50 mL A + ? mL B diluted to 200 mL

pH	7.8	8.0	8.2	8.4	8.6	8.8
?	43.5	41.0	37.7	34.0	29.5	22.0

pH	9.0	9.2	9.4	9.6	9.8	10.0
?	16.7	12.5	8.5	5.7	3.7	2.0

18. Glycine-NaOH Buffer

Stock Solutions

A. 0.2 M solution of glycine (15.01 g in 1000 mL)
B. 0.2 M NaOH (8.0 g of NaOH in 1000 mL)

Formula: 50 mL A + ? mL B diluted to 200 mL

pH	8.6	8.8	9.0	9.2	9.4	9.6	9.8	10.0	10.4	10.6
?	4.0	6.0	8.8	12.0	16.8	22.4	27.2	32.0	38.6	45.5

19. Glycine-NaOH Buffer
Stock Solutions
A. 0.2 M solution of glycine (15.01 g in 1000 mL)
B. 0.2 M NaOH (8.0 g of NaOH in 1000 mL)

Formula: 50 mL A + ? mL B diluted to 200 mL

pH	8.6	8.8	9.0	9.2	9.4	9.6	9.8	10.0	10.4	10.6
?	4.0	6.0	8.8	12.0	16.8	22.4	27.2	32.0	38.6	45.5

20. Borax-NaOH Buffer
Stock Solutions
A. 0.05 M borax solution (19.5 g in 1000 mL; 0.2 M in terms of Na borate)
B. 0.2 M NaOH (8.0 g of NaOH in 1000 mL)

Formula: 50 mL A + ? mL B diluted to 200 mL

pH	9.28	9.35	9.4	9.5	9.6	9.7	9.8	9.9	10.0	10.1
?	0.0	7.0	11.0	17.6	23.0	29.0	34.0	38.6	43.0	46.0

21. Carbonate-Bicarbonate Buffer
Stock Solutions
A. 0.2 M solution of anhydrous sodium carbonate (21.2 g in 1000 mL)
B. 0.2 M $NaHCO_3$ (16.8 g in 1000 mL)

Formula: x mL A + y mL B diluted to 200 mL

pH	9.2	9.3	9.4	9.5	9.6	9.7	9.8	9.9
x	4.0	7.5	9.5	13.0	16.0	19.5	22.0	25.0
y	46.0	42.5	40.5	37.0	34.0	30.5	28.0	25.0

pH	10.0	10.1	10.2	10.3	10.4	10.5	10.6	10.7
x	27.5	30.0	33.0	35.5	38.5	40.5	42.5	45.0
y	22.5	20.0	17.0	14.5	11.5	9.5	7.5	5.0

Useful pH Ranges for Various Buffers

```
pH   1............2............3............4............5............6............7............8............9............10............11
         ------------------  HCL-KCL buffer
                  ------------------  HCL-Glycine buffer
         --------------------  HCL-Phthalate buffer
                  -----------------------------------------  Aconitate buffer
                  ----------------------------------------------  Citrate buffer
                           ------------------------  Acetate buffer
                  -----------------------------------------------------------  Citrate-phosphate buffer
                           ----------------------------------  Succinate buffer
                           --------------------------  Phthalate-NaOH buffer
                                    ---------------------  Maleate buffer
                           -----------------------------  Cacodylate buffer
                                    -----------------------------  Phosphate buffer
    Tris(hydroxymethyl)aminomethane  -------------------------------------------------
     -maleate  buffer
                       Barbital buffer    --------------------------------------
                             Tris buffer    --------------------------
             Boric acid-borax buffer            ---------------------
                   Propanediol buffer      -------------------------------------
                         Glycine buffer        --------------------------
                              Borax-NaOH buffer        -------------
                      Carbonate-bicarbonate buffer        --------------------
pH   1............2............3............4............5............6............7............8............9............10............11
```

Reference

Tiedje, J. M. (1989), personal communication.

3

Standard Oxidation-Reduction Potentials (E_0) at pH 7

Redox Pair	E_0 (mV)
Important Organic Redox Pairs	
Fumarate/succinate	30
Glycine/acetate + NH_4^+	−10
Dihydroxyacetone phosphate/glycerol phosphate	−190
Pyruvate/lactate	−190
CO_2/acetate	−290
CO_2/pyruvate	−310
CO_2/formate	−430
Important Inorganic Redox Pairs	
O_2/H_2O	820
Fe^{3+}/Fe^{2+}	770
NO_3^-/N_2	750
MnO_2/$MnCO_3$	750
NO_3^-/NO_2^-	430
NO_3^-/NH_4^+	380
SO_4^{2-}/HS^-	−220
CO_2/CH_4	−240
S_0/HS^-	−270
H_2O/H_2	−410

Fenchel, T., King, G. M., & Blackburn, T. H. (1998). *Bacterial biogeochemistry*. San Diego, CA: Academic Press.

4

Decay Constants of Radioactive Isotopes

Isotope	Decay Constant λ (days^{-1})	$t_{1/2}$
Calcium-45	0.004	163 days
Carbon-14	**3.3×10^{-7}**	**5760 years**
Cesium-137	5.8×10^{-5}	33 years
Chlorine-36	4.3×10^{-9}	4.4×10^5 years
Chromium-51	0.025	27.8 days
Cobalt-60	3.6×10^{-4}	5.3 years
Copper-64	1.30	12.8 h
Gold-198	0.26	2.69 days
Hydrogen-3	**1.5×10^{-4}**	**12.3 years**
Iodine-125	0.012	60 days
Iodine-131	0.086	8.1 days
Iron-55	0.001	2.9 years
Iron-59	0.015	45.1 days
Lead-210	7.6×10^{-5}	25 years
Manganese-54	0.002	314 days
Mercury-203	0.015	46.6 days
Molybdenum-99	0.252	66 h
Nickel-63	2.2×10^{-5}	85 years
Nitrogen-13	76.8	13 min

(continued)

(Continued)

Isotope	Decay Constant λ (days^{-1})	$t_{1/2}$
Phosphorus-32	**0.048**	**14.3 days**
Phosphorus-33	0.028	25.2 days
Potassium-40	1.5×10^{-12}	1.3×10^9 years
Potassium-42	1.342	12.4 h
Rubidium-86	0.037	18.7 days
Selenium-75	0.005	128 days
Sodium-22	0.001	2.6 years
Strontium-90	6.8×10^{-5}	28 years
Sulfur-35	**0.008**	**87.2 days**
Technetium-99	9.0×10^{-9}	2.1×10^5 years
Zinc-65	0.003	244 days
Zirconium-95	0.011	65 days

Elements in bold are particularly important in soil science studies.

Segel, I. H. (1976). *Biochemical calculations* (2nd ed.). New York: John Wiley & Sons.

APPENDIX

5

Most Probable Number Tables

TABLE A5–1 **Most Probable Numbers to Use with 10-Fold Serial Dilutions and 5 Replications per Dilution**

		Most Probable Number for Observed Value of p_3					
p_1	p_2	0	1	2	3	4	5
0	0	*	0.018	0.036	0.054	0.072	0.090
0	1	0.018	0.036	0.055	0.073	0.091	0.11
0	2	0.037	0.055	0.074	0.092	0.11	0.13
0	3	0.056	0.074	0.093	0.11	0.13	0.15
0	4	0.075	0.094	0.11	0.13	0.15	0.17
0	5	0.094	0.11	0.13	0.15	0.17	0.19
1	0	0.020	0.040	0.060	0.080	0.10	0.12
1	1	0.040	0.061	0.081	0.10	0.12	0.14
1	2	0.061	0.082	0.10	0.12	0.15	0.17
1	3	0.083	0.1	0.13	0.15	0.17	0.19
1	4	0.11	0.13	0.15	0.17	0.19	0.22
1	5	0.13	0.16	0.17	0.19	0.22	0.24
2	0	0.045	0.068	0.091	0.12	0.14	0.16
2	1	0.068	0.092	0.12	0.14	0.17	0.19
2	2	0.093	0.12	0.14	0.17	0.19	0.22
2	3	0.12	0.14	0.17	0.20	0.22	0.25
2	4	0.15	0.17	0.20	0.23	0.25	0.28
2	5	0.17	0.20	0.23	0.26	0.29	0.32
3	0	0.078	0.11	0.13	0.16	0.20	0.23
3	1	0.11	0.14	0.17	0.20	0.23	0.27
3	2	0.14	0.17	0.20	0.24	0.27	0.31

(continued)

TABLE A5–1   (Continued)

		Most Probable Number for Observed Value of p_3					
p_1	p_2	0	1	2	3	4	5
3	3	0.17	0.21	0.24	0.28	0.31	0.35
3	4	0.21	0.24	0.28	0.32	0.36	0.40
3	5	0.25	0.29	0.32	0.37	0.41	0.45
4	0	0.13	0.17	0.21	0.25	0.30	0.36
4	1	0.17	0.21	0.26	0.31	0.36	0.42
4	2	0.22	0.26	0.32	0.38	0.44	0.50
4	3	0.27	0.33	0.39	0.45	0.52	0.59
4	4	0.34	0.40	0.47	0.54	0.62	0.69
4	5	0.41	0.48	0.56	0.64	0.72	0.81
5	0	0.23	0.31	0.43	0.58	0.76	0.95
5	1	0.33	0.46	0.64	0.84	1.10	1.30
5	2	0.49	0.70	0.95	1.20	1.50	1.80
5	3	0.79	1.10	1.40	1.80	2.10	2.50
5	4	1.30	1.70	2.20	2.80	3.50	4.30
5	5	2.40	3.50	5.40	9.20	16.0	*

Alexander, M. (1982). Most probable number methods for microbial populations. In A. L. Page et al. (Eds.), *Methods of soil analysis, Part 2. Microbiological and biochemical properties* (2nd ed., pp. 815–820). Madison, WI: Soil Science Society of America.

TABLE A5–2   Factors for the 95 Percent Confidence Interval for MPN Analysis

	Dilution Ratio			
Number of Replicate Samples	2	4	5	10
1	4.00	7.14	8.32	14.45
2	2.67	4.00	4.47	6.61
3	2.23	3.10	3.39	4.68
4	2.00	2.68	2.88	3.80
5	1.86	2.41	2.58	3.30
6	1.76	2.23	2.38	2.98
7	1.69	2.10	2.23	2.74
8	1.64	2.00	2.12	2.57
9	1.58	1.92	2.02	2.43
10	1.55	1.86	1.95	2.32

Alexander, M. (1982). Most probable number methods for microbial populations. In A. L. Page et al. (Eds.), *Methods of soil analysis, Part 2. Microbiological and biochemical properties* (2nd ed., pp. 815–820). Madison, WI: Soil Science Society of America.

6

Sample Microsoft Excel Program for Determining MPN

The following program should allow you to calculate MPN for any system employing 10-fold dilutions and five replicates per dilution:

Step 1. Headings

Start dilution	1	2	3	4	5	6	7	$p_1p_2p_3$	MPN	MPN bact.
0	0	0	0	0	0	0	0	0	0	0.00E + 00
0	0	0	0	0	0	0	0	0	0	0.00E + 00

Step 2. Enter Data Starting in Cell A2

A. Enter the starting dilution factor in Column A
B. Enter the number of positive results (or negative results if you are looking at absence) in each cell corresponding to dilutions 1 to 7.

Step 3. Insert the Following Equation to Determine p_1, p_2, and p_3:

=IF(F3=MAX(B3:F3),F3*100+G3*10+H3, IF(E3=MAX(B3:F3),E3*100
+F3*10+G3,IF(D3=MAX(B3:F3),D3*100+E3*10+F3,(IF(C3=
MAX(B3:F3),C3*100+D3*10+E3,B3*100+C3*10+D3)))))

Step 4. Insert the Following Equation to Determine MPN:

=VLOOKUP(I3,'MPN Table'!A$2:B$217,2)

Step 5. Insert the Following Equation to Determine MPN Bacteria:

=IF(F19=MAX(B19:F19),J19*10^(5+A19),IF(E19=MAX(B19:F19),
J19*10^(4+A19),IF(D19=MAX(B19:F19),J19*10^(3+A19),IF(C19=MAX
(B19:F19),J19*10^(2+A19),J19*10^(1+A19)))))

Step 6. Enter the Following Data to Create the Table for MPN Values:

0		22	0.074	44	0.15	110	0.04
1	0.018	23	0.092	45	0.17	111	0.061
2	0.036	24	0.11	50	0.094	112	0.081
3	0.054	25	0.13	51	0.11	113	0.1
4	0.072	30	0.56	52	0.13	114	0.12
5	0.09	31	0.074	53	0.15	115	0.14
10	0.018	32	0.093	54	0.17	120	0.061
11	0.036	33	0.11	55	0.19	121	0.082
12	0.055	34	0.13	100	0.02	122	0.1
13	0.073	35	0.15	101	0.04	123	0.12
14	0.091	40	0.075	102	0.06	124	0.15
15	0.11	41	0.094	103	0.08	125	0.17
20	0.037	42	0.11	104	0.1	130	0.083
21	0.055	43	0.13	105	0.12	131	0.1

132	0.13	240	0.15	344	0.36	452	0.56
133	0.15	241	0.17	345	0.4	453	0.64
134	0.17	242	0.2	350	0.25	454	0.72
135	0.19	243	0.23	351	0.29	455	0.81
140	0.11	244	0.25	352	0.32	500	0.23
141	0.13	245	0.28	353	0.37	501	0.31
142	0.15	250	0.17	354	0.41	502	0.43
143	0.17	251	0.2	355	0.45	503	0.58
144	0.19	252	0.23	400	0.13	504	0.76
145	0.22	253	0.26	401	0.17	505	0.95
150	0.13	254	0.29	402	0.21	510	0.33
151	0.16	255	0.32	403	0.25	511	0.46
152	0.17	300	0.078	404	0.3	512	0.64
153	0.19	301	0.11	405	0.36	513	0.84
154	0.22	302	0.13	410	0.17	514	1.1
155	0.24	303	0.16	411	0.21	515	1.3
200	0.045	304	0.2	412	0.26	520	0.49
201	0.068	305	0.23	413	0.31	521	0.7
202	0.091	310	0.11	414	0.36	522	0.95
203	0.12	311	0.14	415	0.42	523	1.2
204	0.14	312	0.17	420	0.22	524	1.5
205	0.16	313	0.2	421	0.26	525	1.8
210	0.068	314	0.23	422	0.32	530	0.79
211	0.092	315	0.27	423	0.38	531	1.1
212	0.12	320	0.14	424	0.44	532	1.4
213	0.14	321	0.17	425	0.5	533	1.8
214	0.17	322	0.2	430	0.27	534	2.1
215	0.19	323	0.24	431	0.33	535	2.5
220	0.093	324	0.27	432	0.39	540	1.3
221	0.12	325	0.31	433	0.45	541	1.7
222	0.14	330	0.17	434	0.52	542	2.2
223	0.17	331	0.21	435	0.59	543	2.8
224	0.19	332	0.24	440	0.34	544	3.5
225	0.22	333	0.28	441	0.4	545	4.3
230	0.12	334	0.31	442	0.47	550	2.4
231	0.14	335	0.35	443	0.54	551	3.5
232	0.17	340	0.21	444	0.62	552	5.4
233	0.2	341	0.24	445	0.69	553	9.2
234	0.22	342	0.28	450	0.41	554	16
235	0.25	343	0.32	451	0.48	555	

7

First-Order Mineralization Rate Constants for Various Organic Materials

TABLE A7–1 Decomposition Rates for Plant Residues, Plant Compounds, Manures, and Soil Organic Matter

Substrate	k (day^{-1})	Reference
Wheat straw		
Fast pool	0.019	
Slow pool	7.1×10^{-4}	
Wheat straw		
Sugars	0.2	Paul & Clark, 1996
Cellulose/Hemicellulose	0.08	
Lignin	0.01	
Rye straw	0.01–0.04	Paul & Clark, 1996
Rye straw	0.02	Voroney et al., 1981
Corn residue	0.011–0.056	
Native grass	0.006	Voroney et al., 1981
Sugars-Amino acids	0.2	Voroney et al., 1981
Cellulose-Hemicellulose	0.08	Voroney et al., 1981
Hemicellulose	0.03	Voroney et al., 1981
Lignin	0.003–0.01	Voroney et al., 1981
Microbial biomass		Voroney et al., 1981
Fast pool	0.8	
Slow pool	0.3	

(continued)

TABLE A7–1 (Continued)

Substrate	k (day^{-1})	Reference
Fungal cytoplasm	0.04	Voroney et al., 1981
Fungal cell wall	0.02	Voroney et al., 1981
Poultry litter N		
Fast pool	2.6	
Slow pool	0.041	
Biosolids		
Fast pool	0.165	Gilmour et al., 1996
Slow pool	2.64×10^{-3}	
Organic matter		
Active pool	3×10^{-4}	Paul & Clark, 1996
Old/Passive pool	8×10^{-7}	

Gilmour, J. T., Clark, M. D., & Daniel, S. M. (1996). *Journal of Environmental Quality, 25,* 766–770.
Paul, E. A., & Clark, F. E. (1996). *Soil microbiology and biochemistry.* San Diego, CA: Academic Press.
Voroney, R. P., Van Veen, J. A., & Paul, E. A. (1981). *Journal of Soil Science, 61,* 211–224.

TABLE A7–2 Decomposition of Organic Compounds in Laboratory Conditions[a]

Substrate	Decomposition Rate (day^{-1})		Relative Decomposition Rate Week 1 (Glycine = 1)	% Decomposed (28 weeks)
	Week 1	Week 4–28		
Simple Organic Compounds				
Glycine	0.192	0.004	1.00	91
Glucose	0.187	0.003	0.97	90
Benzoic acid	0.163	0.002	0.85	84
Cysteine	0.111	0.002	0.58	78
Caffeic acid	0.073	0.001	0.38	68
2,4-D side chain carbon	0.028	0.004	0.15	88
Catechol	0.017	6.0×10^{-4}	0.09	26
2,4-D ring carbon	0.009	0.002	0.05	75
Simple Plant Polymers				
Starch	0.093	0.004	0.48	86
Casein	0.124	0.003	0.65	85
Cellulose	0.005	0.006	0.03	84

(continued)

TABLE A7–2   (Continued)

Substrate	Decomposition Rate (day^{-1})		Relative Decomposition Rate Week 1 (Glycine = 1)	% Decomposed (28 weeks)
	Week 1	Week 4–28		
Microbial Biomass				
Arthrobacter cells	0.131	0.003	0.68	87
Earthworms	0.127	0.003	0.66	84
Penicillium cells	0.117	0.002	0.61	79
Azotobacter chroococcum	0.134	0.002	0.70	78
Azotobacter polysaccharide	0.004	0.005	0.02	68
Aspergillus glaucus	0.043	0.001	0.22	54
Aspergillus glaucus melanin	0.004	2.5×10^{-4}	0.02	10
Plant Residue/Manure				
Lima bean straw	0.064	0.004	0.33	79
Wheat straw	0.043	0.003	0.22	67
Sudan grass	0.059	0.002	0.31	66
Cow manure	0.028	0.002	0.15	50
Peat moss	0.001	9.6×10^{-4}	0.005	17
Wood Products				
Walnut wood	0.010	0.002	0.05	53
Almond shells	0.018	0.001	0.09	41
Douglas fir wood	0.003	0.002	0.02	34
Ponderosa pine needles	0.018	7.2×10^{-4}	0.09	32
Incense cedar wood	0.001	1.3×10^{-4}	0.005	3
Soil Organic Matter				
Soil humic acid	0.001	5.4×10^{-5}	0.005	2

[a]Percent decomposition measured as CO_2 evolved at 22°C in a Greenfield sandy loam soil.

Martin, J. P., & Focht, D. D. (1977). *Biological properties of soils.* In L. F. Elliot & F. J. Stevenson (Eds.), *Soils for management of organic wastes and waste waters* (pp. 115–169). Madison, WI: Soil Science Society of America.

8

Bunsen Absorption Coefficients for Various Gases in Water

Temp.	Bunsen Absorption Coefficients (α)				
°C	N$_2$O	NO	N$_2$	C$_2$H$_2$	O$_2$
5	1.06	6.46×10^{-2}	2.11×10^{-2}	1.51	4.30×10^{-2}
10	0.882	5.75×10^{-2}	1.89×10^{-2}	1.31	3.82×10^{-2}
15	0.743	5.17×10^{-2}	1.71×10^{-2}	1.16	3.43×10^{-2}
20	0.632	4.70×10^{-2}	1.57×10^{-2}	1.03	3.11×10^{2}
25	0.544	4.31×10^{-2}	1.45×10^{-2}	0.928	2.85×10^{-2}
30	0.472	3.99×10^{-2}	1.36×10^{-2}	0.843	2.64×10^{-2}
35	0.414	3.72×10^{-2}	1.28×10^{-2}	0.771	2.46×10^{-2}

9

Significant Elements in Soil

TABLE A9–1 Atomic Number and Molecular Weights of Significant Elements in Soil[a]

Element	Symbol	Atomic Number	Atomic Weight
Aluminum	Al	13	26.97
Antimony	Sb	51	121.76
Argon	A	18	39.94
Arsenic	As	33	74.92
Barium	Ba	56	137.36
Beryllium	Be	4	9.02
Boron	B	5	10.82
Bromine	Br	35	79.92
Cadmium	Cd	48	112.41
Calcium	Ca	20	40.08
Carbon	C	6	12.01
Chlorine	Cl	17	35.46
Chromium	Cr	24	52.01
Cobalt	Co	27	58.94
Copper	Cu	29	63.57
Fluorine	F	9	19.00
Gallium	Ga	31	69.72
Germanium	Ge	32	72.61
Gold	Au	79	197.20

(*continued*)

TABLE A9–1   (Continued)

Element	Symbol	Atomic Number	Atomic Weight
Helium	He	2	4.00
Hydrogen	H	1	1.01
Indium	In	49	114.82
Iodine	I	53	126.92
Iron	Fe	26	55.84
Krypton	Kr	36	83.80
Lead	Pb	82	207.21
Lithium	Li	3	6.941
Magnesium	Mg	12	24.32
Manganese	Mn	25	54.93
Mercury	Hg	80	200.61
Molybdenum	Mo	42	95.95
Neon	Ne	10	20.18
Nickel	Ni	28	58.69
Nitrogen	N	7	14.01
Nobelium	Nb	41	92.91
Oxygen	O	8	16.00
Palladium	Pd	46	106.42
Phosphorus	P	15	30.98
Platinum	Pt	78	195.08
Potassium	K	19	39.10
Radon	Rn	86	222.00
Rubidium	Rb	37	85.47
Ruthenium	Ru	44	101.07
Scandium	Sc	21	44.957
Selenium	Se	34	78.96
Silicon	Si	14	28.06
Silver	Ag	47	107.88
Sodium	Na	11	23.00
Strontium	Sr	38	87.62
Sulfur	S	16	32.06
Technetium	Tc	43	98.91
Tellurium	Te	52	127.62
Tin	Sn	50	118.70

(*continued*)

TABLE A9–1 (Continued)

Element	Symbol	Atomic Number	Atomic Weight
Titanium	Ti	22	47.90
Tungsten	W	74	183.92
Uranium	U	92	238.07
Vanadium	V	23	50.95
Xenon	Xe	54	131.29
Yttrium	Y	39	88.91
Zinc	Zn	30	65.38
Zirconium	Zr	40	91.22

[a]Only those elements with an atomic number <53 play a biological role in soil.

TABLE A9–2 Equivalent Weights of Significant Cations and Anions in Soil

Element	Symbol	Typical Charge	Atomic Weight	Equivalent Weight at Typical Charge
Arsenic	As	-2 (as AsO_4^{2-})	74.92	37.46
Barium	Ba	$+2$	137.36	68.68
Bromine	Br	-1	79.92	79.92
Cadmium	Cd	$+2$	112.41	56.20
Calcium	Ca	$+2$	40.08	20.04
Carbon	C	-2 (as CO_3^{2-})	12.01	6.00
Chlorine	Cl	-1	35.46	35.46
Copper	Cu	$+2$	63.57	31.78
Fluorine	F	-1	19.00	19.00
Hydrogen	H	$+1$	1.01	1.01
Lead	Pb	$+2$	207.21	103.60
Magnesium	Mg	$+2$	24.32	12.16
Mercury	Hg	$+2$	200.61	100.30
Nickel	Ni	$+2$	58.69	29.34
Nitrogen	N	$+1$ (as NH_4^+)	14.01	14.01
		-1 (as NO_3^-)		

(continued)

TABLE A9–2 (Continued)

Element	Symbol	Typical Charge	Atomic Weight	Equivalent Weight at Typical Charge
Potassium	K	+1	39.10	39.10
Silver	Ag	+2	107.88	53.94
Sodium	Na	+1	23.00	23.00
Sulfur	S	−2 (as SO_4^{2-})	32.07	16.04
Zinc	Zn	+2	65.38	32.69

TABLE A9–3 Relative Abundance of Elements by Weight

Whole Earth Abundance		Earth Crust Abundance	
Element	%	Element	%
Iron (Fe)	35	Oxygen (O)	46
Oxygen (O)	30	Silicon (Si)	28
Silicon (Si)	15	Aluminum (Al)	8
Magnesium (Mg)	13	Iron (Fe)	6
Nickel (Ni)	2.4	Magnesium (Mg)	4
Sulfur (S)	1.9	Calcium (Ca)	2.4
Calcium (Ca)	1.1	Potassium (K)	2.3
Aluminum (Al)	1.1	Sodium (Na)	2.1
Other	<1	Other	<1

Schlesinger, W. H. (1997). *Biogeochemistry: An analysis of global change.* San Diego, CA: Academic Press.

10

Properties of Concentrated Reagent Acids and NH$_3$ with Dilution Directions to Make One Normal Solution

Reagent Name	Formula	Mol. Wt. (g)	Specific Gravity (g/mL)	Percent (w/w)	g/100 mL	Approximate Normality (eq/L)	Amt. Required for 1 L of 1 N Solution (mL)
Acids							
Acetic acid	CH_3COOH	60.0	1.06	100	106.0	17.4	57.5
Formic acid	HCOOH	46.0	1.20	88	105.6	24.0	41.6
Hydrochloric acid	HCl	36.5	1.19	37	44.0	12.1	82.5
Nitric acid	HNO_3	63.0	1.42	70	91.0	15.8	63.5
Phosphoric acid	H_3PO_4	98.0	1.71	85	146.0	44.5	22.5
Perchloric acid	$HClO_4$	100.5	1.66	70	116.2	11.6	86.5
Sulfuric acid	H_2SO_4	98.1	1.84	96	173.0	35.2	29.0
Bases							
Ammonia	NH_3	17.0	0.91	28	22.8	14.8	67.5

Segel, I. H. (1976). *Biochemical calculations.* New York: John Wiley & Sons.

11

Microsoft Excel Tips and Pointers

R. Barnhisel, University of Kentucky
J. M. Thompson, West Virginia University

1. To perform basic mathematical operations.

addition:	=B13+2	or	=B13+B14
subtraction:	=B13-2	or	=B13-B14
multiplication:	=B13*2	or	=B13*B14
division:	=B13/2	or	=B13/B14
exponentiation:	=B13^2	or	=B13^B14

If you combine several operations in a single formula, Excel performs the operations in the following order:

(1) exponentiation
(2) multiplication and division
(3) addition and subtraction

To change the order of evaluation, enclose the part of the formula to be calculated first in parentheses. For example:

$$(a) = 5*2^3-1 = 39$$

$$(b) = (5*(2^3))-1 = 39$$

$$(c) = 5*((2^3)-1) = 35$$

$$(d) = 5*2^(3-1) = 20$$

$$(e) = (5*2)^(3-1) = 100$$

$$(f) = (5*2)^3-1 = 999$$

2. To add the values within a column or block of cells.

$$=sum(B13:B21)$$

3. To calculate the mean from a set of data.

$$=average(B13:B21)$$

4. To calculate standard deviation from a set of data.

$$=stdev(B13:B21)$$

5. To determine the probability of significance using a *t*-test—this test is used to determine the *similarity* between two sets of data (i.e., are their means equal?).

$$=ttest(B13:B21,C13:C21,2,1)$$

The first two groups (B13:B21 and C13:C21) are the data sets you are comparing. The third value indicates the number of tails (1 = one tailed, 2 = two-tailed)—for a test of equal means; a two-tailed *t*-test is selected in this example. The fourth value indicates the type of *t*-test (1 = paired, 2 = equal variance, 3 = unequal variance); in this example, a test of paired means is made. If the resulting probability is >0.05, then the data sets are not significantly different, if <0.05, the data sets are different. In this case, 0.05 refers to the **alpha value** or **significance value.**

6. To determine a simple linear regression relationship between two sets of data, you compare the slope of a line between two columns of data (the first set is the dependent variable).

$$=slope(B13:B21,C13:C21)$$

7. To get the *Y*-intercept of a line between two columns (the first set is the dependent variable).

$$=intercept(B13:B21,C13:C21)$$

The resulting equation for a line is

$$Y = a + bX$$

where

Y = dependent variable
a = Y-intercept
b = slope
X = independent variable

8. To determine the coefficient of correlation between the two variables.

$$=correl(B13:B21,C13:C21)$$

For any other functions, go into the HELP menu, or select the FUNCTION command under INSERT to view other possible functions.

Index

CPSIA information can be obtained
at www.ICGtesting.com
Printed in the USA
FFOW04n1953110713
1390FF